New developments in ESP teaching and learning research

Edited by Cédric Sarré and Shona Whyte

Published by Research-publishing.net, not-for-profit association
Voillans, France, info@research-publishing.net

© 2017 by Editors (collective work)
© 2017 by Authors (individual work)

New developments in ESP teaching and learning research
Edited by Cédric Sarré and Shona Whyte

Rights: This volume is published under the Attribution-NonCommercial-NoDerivatives International (CC BY-NC-ND) licence; **individual articles may have a different licence**. Under the CC BY-NC-ND licence, the volume is freely available online (https://doi.org/10.14705/rpnet.2017.cssw2017.9782490057016) for anybody to read, download, copy, and redistribute provided that the author(s), editorial team, and publisher are properly cited. Commercial use and derivative works are, however, not permitted.

Disclaimer: Research-publishing.net does not take any responsibility for the content of the pages written by the authors of this book. The authors have recognised that the work described was not published before, or that it was not under consideration for publication elsewhere. While the information in this book is believed to be true and accurate on the date of its going to press, neither the editorial team, nor the publisher can accept any legal responsibility for any errors or omissions that may be made. The publisher makes no warranty, expressed or implied, with respect to the material contained herein. While Research-publishing.net is committed to publishing works of integrity, the words are the authors' alone.

Trademark notice: product or corporate names may be trademarks or registered trademarks, and are used only for identification and explanation without intent to infringe.

Copyrighted material: every effort has been made by the editorial team to trace copyright holders and to obtain their permission for the use of copyrighted material in this book. In the event of errors or omissions, please notify the publisher of any corrections that will need to be incorporated in future editions of this book.

Typeset by Research-publishing.net
Cover design and cover photos by © Raphaël Savina (raphael@savina.net)

ISBN13: 978-2-490057-01-6 (Ebook, PDF, colour)
ISBN13: 978-2-490057-02-3 (Ebook, EPUB, colour)
ISBN13: 978-2-490057-00-9 (Paperback - Print on demand, black and white)
Print on demand technology is a high-quality, innovative and ecological printing method; with which the book is never 'out of stock' or 'out of print'.

British Library Cataloguing-in-Publication Data.
A cataloguing record for this book is available from the British Library.

Legal deposit, France: Bibliothèque Nationale de France - Dépôt légal: décembre 2017.

Table of contents

v Notes on contributors

xiii Acknowledgements

xv Foreword
Dan Douglas

1 Introduction to new developments in ESP teaching and learning research
Shona Whyte and Cédric Sarré

Section 1.

Laying the groundwork: needs analysis, programme design, and course development

15 Using video materials in English for technical sciences: a case study
Danica Milosevic

31 Dynamic and complex system approach to needs analysis, course development and evaluation of LANSOD courses in a French musicology undergraduate programme
Aude Labetoulle

51 Designing and implementing ESP courses in French higher education: a case study
Susan Birch-Bécaas and Laüra Hoskins

Section 2.

Building confidence: addressing particular difficulties

73 Towards a dynamic approach to analysing student motivation in ESP courses
Daniel Schug and Gwen Le Cor

Table of contents

93 Teaching compound nouns in ESP: insights from cognitive semantics
Marie-Hélène Fries

109 When storytelling meets active learning: an academic reading experiment with French MA students
Pauline Beaupoil-Hourdel, Hélène Josse, Loulou Kosmala, Katy Masuga, and Aliyah Morgenstern

Section 3.

Moving ahead: towards new practices

133 The SocWoC corpus: compiling and exploiting ESP material for undergraduate social workers
Jane Helen Johnson

153 Identifying and responding to learner needs at the medical faculty: the use of audio-visual specialised fiction (FASP)
Rebecca Franklin-Landi

171 The effect of form-focussed pre-task activities on accuracy in L2 production in an ESP course in French higher education
Rebecca Starkey-Perret, Sophie Belan, Thi Phuong Lê Ngo, and Guillaume Rialland

196 Author index

Notes on contributors

Editors

Cédric Sarré is a senior lecturer at Université Paris-Sorbonne, School of Education (ESPE de Paris) where he teaches English as a Foreign Language (EFL) and English language learning and teaching. He is a member of the CeLiSo (Centre de Linguistique en Sorbonne, EA 7332) research unit. His research focusses on language learning technologies – Computer-Assisted Language Learning (CALL) and Computer-Mediated Communication (CMC). His research interests include ESP didactics, ESP course development in online settings, language proficiency testing, peer interaction in L2 learning, and teacher education. With Shona Whyte, he co-chairs the DidASP special interest group on ESP didactics within GERAS (Groupe d'Etude et de Recherche en Anglais de Spécialité), the French ESP research association.

Shona Whyte is Associate Professor of English at the Université Côte d'Azur where she teaches English as a Foreign Language (EFL), translation, and second language (L2) studies. Her research interests within Computer-Assisted Language Learning (CALL) include classroom interaction and teacher integration of technologies. With Cédric Sarré, she co-directs a special interest group within the French association GERAS on ESP didactics, that is, research in teaching English for Specific Purposes (ESP). She is also interested in research cultures in L2 studies and language didactics and reviews for English and French language journals in CALL and language education.

Authors

Pauline Beaupoil-Hourdel is a postdoctoral fellow in the PRISMES Lab (EA 4398) at Sorbonne Nouvelle University. She received her doctorate in linguistics, first language acquisition, and multimodal analyses in natural interactions. Her recent research focusses on the relationship between language learning and innovative approaches to teaching, as well as on the role of multimodal communication in language acquisition and production.

Notes on contributors

Sophie Belan is an associate professor and course coordinator at the Foreign Languages and International Trade (LEA) department of the University of Nantes, France, where she teaches business English. Her research focusses on English for specific purposes, task-based learning and teaching, blended learning, and CALL. She is a member of CRINI (Centre de Recherche sur les Identités Nationales et l'Interculturalité, EA 1162) and of GERAS (Groupe d'Étude et de Recherche en Anglais de Spécialité).

Susan Birch-Bécaas is a senior lecturer at the University of Bordeaux, where she is head of the language department for students in the health sciences and human sciences. She coordinates ESP courses for students of public health and dentistry, and writing courses for doctoral students. Her research interests include the analysis of scientific discourse and its pedagogical applications, especially for English for research and publication purposes. She has published on the linguistic needs of French researchers publishing in English. She also works with the DLC (Département de Langues et Cultures) team on the 'Teaching Academic Content in English' course for Défi international.

Dan Douglas, Professor Emeritus in the applied linguistics programme at Iowa State University, has been working in language for specific purposes since his student days at Edinburgh University in the early 1970's. He has taught and conducted research at the University of Khartoum, Hiroshima University, the University of Michigan, and Iowa State. His books include Assessing Languages for Specific Purposes (Cambridge, 2000), Assessing Language through Computer Technology (with Carol Chapelle, Cambridge, 2006), and Understanding Language Testing (Routledge, 2010). His current interests include English for aviation, and English for nursing.

Rebecca Franklin-Landi has a PhD in American cinema and civilisation studies, and currently teaches at the medical faculty of Nice University, France. She has published on the role of the foreigner in Hollywood cinema. Since then, her research has concentrated on the use of film in English for medical purposes teaching. She has also carried out dictogloss research with

Notes on contributors

her students within a wider English for specific purposes framework. At present, her research focusses on the acquisition of EMP oral skills.

Marie-Hélène Fries is an English lecturer in Université Grenoble-Alpes (France), and is part of the LCEA4 laboratory. She teaches English to chemistry and process engineering students. Her research interests stem from the intuition that a cognitive view of language can be an entrance door into specialised fields in science and technology. This includes both using metaphorical terms and blending to understand globally what is at stake in scientific domains and reflecting on the relevance of cognitive semantics for science and technology students.

Laüra Hoskins is an ESP teacher at the University of Bordeaux where she coordinates ESP courses for students of dentistry, psychology, and sociology. Her current interests are in blended learning and she has designed and implemented a Moodle-based blended learning course for a cohort of 1050 first-year psychology, sociology, and sports students. She also has experience teaching students of biology and medicine and more recently in teaching aspects of scientific communication to doctoral students. With her DLC colleagues, she contributes to the course 'Teaching Academic Content in English' for Défi international and has communicated at Ranacles on the modules she co-designed.

Jane Helen Johnson is Research Assistant at the Department of Modern Languages, Literatures and Cultures of the University of Bologna. A qualified teacher of English as a foreign language, she currently teaches English for social workers at the School of Political Sciences as well as English for specific purposes on the Master's degree course in language, society and communication. Her research interests include English as a medium of instruction, translation, and Corpus-Assisted Discourse Studies (CADS), while her current research projects combine CADS and English for specific purposes.

Hélène Josse (PhD, English Linguistics) is a senior lecturer at Sorbonne Nouvelle University, Paris, France. Her research focusses on active, interactive pedagogy at the university with a special interest in didactics of English grammar for non-native English speakers. She is a member of the PRISMES (EA 4398) lab and the team SeSyLIA (Semantics and Syntax - Language In Action).

Loulou Kosmala is a research assistant within the PRISMES research unit at Sorbonne Nouvelle University. She recently obtained her Master's degree in English linguistics.

Aude Labetoulle is a third year PhD student co-directed by Annick Rivens Mompean and Jean-Claude Bertin at the University of Lille SHS. She was trained at the École Normale Supérieure (ENS) de Cachan and she holds an 'agrégation' in English. She taught English for specific purposes at the ENS de Cachan, at the University of Paris Diderot Institute of Technology (Physical Measurements), and currently teaches at the University of Lille SHS. She investigates specialised forms of English and LANSAD teaching.

Gwen Le Cor is a full professor at the Université Paris 8 where she is also co-director of the 'Transferts critiques anglophones' research lab. Her current research interests include the intersections between literature and sciences, English for scientific purposes, and contemporary American literature. She published extensively on the print and digital works of Jen Bervin, Percival Everett, Jonathan Safran Foer, Nick Monfort, Flannery O'Connor, Art Spiegelman, Stephanie Strickland, Steve Tomasula, and Robert Penn Warren. Her work has also focussed on intersemiocity, scientific discourse, literature-computer science nodes, and subjects pertaining to the digital humanities.

Thi Phuong Lê Ngo is a PhD student in English teaching at the University of Nantes. Her doctoral thesis focusses on the development and evaluation of the virtual resource centre for the business English programme for first-year students. Before coming to Nantes, she was a teacher of English at the National University of Vietnam.

Katy Masuga holds a PhD in comparative literature and a joint PhD in literary theory and criticism, with a research focus on intersections between the sciences and humanities. Her publications include two monographs on Henry

Miller, the novels The Origin of Vermilion (2016) and The Blue of Night (2017), and numerous short stories, articles, and essays. Fellow at Sorbonne Nouvelle University, Masuga teaches for the University of Washington (Seattle) programme in Paris.

Danica Milosevic has taught technical English courses for the past seven years in modern computer technologies, communication technologies, industrial engineering, road traffic, environmental protection and civil engineering at the College of Applied Technical Sciences Nis in Serbia. Her focus is on methodology in the ESP environment. She has completed postgraduate studies in Anglo-American literature and culture at the Faculty of Philosophy, as well as her third year of doctoral studies. She is currently a doctoral candidate in English literature at the Faculty of Philology and Arts in Kragujevac, Serbia.

Aliyah Morgenstern is a professor at Sorbonne Nouvelle University where she teaches English linguistics, language acquisition, and multimodal interaction. She is dean of the Graduate School in English, German, and European Studies. She is the scientific coordinator of the Research and Storytelling project described in her paper. Her major publications focus on language acquisition using socio-pragmatic, constructionist, and functionalist perspectives. In collaboration with other linguists, psychologists, speech therapists, and anthropologists, she analyses language use in context with a multimodal perspective. Her aim is to combine what we learn from gestures, phonology, morpho-syntax, and discourse to understand the blossoming of our pluri-semiotic linguistic skills.

Guillaume Rialland obtained his Master's degree in foreign language teaching from the University of Nantes in 2014. He has since then extended his collaboration and research with his colleagues while teaching business English and translation in the department of Foreign Languages and International Trade (LEA) at the University of Nantes.

Daniel Schug is a second-year PhD student completing a joint-PhD with the Università Ca' Foscari Venezia and the Université Paris 8. He has completed

Masters' degrees in language teaching in both the USA and France, and has taught English and French in both countries at various levels. His current interests include motivation in university language courses as well as the language learning habits of students studying abroad. For his doctoral research, he is seeking to use the complex dynamic systems theory to analyse motivation in university courses of English for specific purposes.

Rebecca Starkey-Perret is an associate professor at the School of Education of Nantes (ESPE) and head of the programme for the Master in English teaching. She continues to teach in the Foreign Languages and International Trade (LEA) department of the University of Nantes. Her research themes include teacher and student representations, plurilingualism, task-based language learning, and CALL. She is a member of CRINI (Centre de Recherche sur les Identités Nationales et l'Interculturalité, EA 1162).

Reviewers

Jean-Claude Bertin is Professor Emeritus of English language learning and teaching at the University of Le Havre Normandie, France. He has published a large number of articles and reports in the field of Computer-Assisted Language Learning (CALL) and distance learning, in national as well as international journals, and authored or co-authored numerous books (Des outils pour des langues – multimédia et apprentissage, 2001; Second-language distance learning and teaching: theoretical perspectives and didactic ergonomics, 2010; L'homme @ distance, regards croisés – innovation et développement, 2013). From 2011 to 2014, he was President of the French ESP research association GERAS (Groupe d'Étude et de Recherche en Anglais de Spécialité) and Director of ASp journal.

Dacia Dressen-Hammouda is Associate Professor of ESP and technical communication at Université Clermont Auvergne, France, where she directs the Bachelors' and Masters' programmes in information design and multilingual technical documentation. Her current research interests include L1 and L2 writing research, indexicality, genre theory, intercultural rhetoric, and the

Notes on contributors

discoursal construction of expertise in novice and non-native disciplinary writers. Her forthcoming book is 'Learning the genres of geology: the role of indexicality and agency in the emergence of situated writing expertise.'

Dan Frost studied languages and linguistics at York University in the UK, and at Strasbourg, Aix-en-Provence, and Bordeaux universities in France. His doctorate is in teaching pronunciation for learners of English for specific and academic purposes using technology. He taught English in the UK, Thailand, and Sweden before settling in France where, after teaching English in secondary schools, in an IT department, and in Applied Foreign Languages, he is now a senior lecturer in the Lifelong Learning department at Grenoble Alpes University, France. His main research interests are teaching pronunciation, oral English, and computer-mediated learning.

Alice Henderson is an associate professor of English at the Université Savoie, Mont Blanc in Chambéry, France, where for the past 23 years she has taught students (primarily those majoring in English or in the social sciences) as well as university professors who want to teach their subject in English. Her research is focussed on non-native spoken English, leading her to work on the didactics of spoken English, the teaching and learning of English pronunciation, and language policy, especially CLIL (Content-Language Integrated Learning). She has been involved in teacher training in several European countries and the United States.

Monique Mémet is Associate Professor Emerita at the École Normale Supérieure Paris-Saclay, France, where she taught theory and practice of English for specific purposes to undergraduate and graduate students. She co-edited 'L'anglais de spécialité en France' (Bordeaux: GERAS, 2001) and published numerous articles in national and international journals. She is the editor of the electronic version of ASp.

Annick Rivens Mompean is Professor of English didactics at Lille SHS University (France) where she is currently Vice-President in charge of language policy. Her research interests include Computer-Assisted Language Learning (CALL) and Computer-Mediated Communication (CMC). She is also interested in the development of autonomy related to language learning, especially in the learning context of language resource centres. She is the president of RANACLES, the French association of language centres in higher education affiliated to the European Confederation of Language Centres in Higher Education (CERCLES).

Geoff Sockett is Professor of linguistics at Paris Descartes University (Paris, France) where he is also currently Vice-President for international relations. His research interests focus on online language learning in informal contexts and more particularly incidental acquisition from leisure activities, such as watching videos and listening to music. He has published research in these fields in both French and English, framing such acquisition in cognitive linguistics and complex systems theories. He is a reviewer for a number of journals, including ReCall and Recherches en Didactique des Langues et des Cultures (RDLC).

Acknowledgements

We would like to express our appreciation to the authors of the chapters in this book for their valuable contributions, to the members of our editorial board who generously gave their time and expertise to review them, and to Professor Dan Douglas for accepting our invitation to write a foreword to this collection.

We are grateful to a number of institutions and organisations for their sponsorship of this project: the Université Paris-Sorbonne, especially the CeLiSo (Centre de Linguistique en Sorbonne, EA7332), the Ecole Doctorale V – Concepts et Langages – and the ESPE de l'Académie de Paris, as well as the Université Côte d'Azur. We owe a special debt of gratitude to the French association for ESP in higher education, the GERAS (Groupe d'Etude et de Recherche en Anglais de Spécialité) for encouragement as well as financial support.

Indeed, the present volume is at least in part a tribute to the work of many colleagues, ESP practitioners, and researchers who have participated in workshops, SIG meetings, and seminars, including GERAS and ESSE conferences in recent years. Their contributions through presentations and discussions have helped us clarify our perceptions of the field and shape many of the ideas in this book.

Last, but not least, we express our thanks to Research-publishing.net – and Sylvie Thouësny in particular – for their unfailing patience, insight, and professionalism which smoothed our path at every turn. We are delighted to participate in this kind of open approach to academic publishing and warmly recommend the team.

Cédric Sarré and Shona Whyte

Foreword

Dan Douglas[1]

John Swales (1985), in his pioneering book, *Episodes in ESP*, makes the following observation:

> "My own experience of [...] ESP [...] is that the profession as a whole, and with all too few exceptions, operates within the "here and now" of their actual teaching situation [...]. ESP practitioners tend not to look across to other ESP situations and to other ESP endeavors, whether similar or dissimilar to their own, to see what lessons may be learnt, what insights might be gained, or what useful short-cuts can be made" (p. ix).

In the present volume, the editors have put together a number of research reports from very specific ESP contexts that respond to Swales's (1985) point wonderfully. The range of ESP situations presented in the book is quite broad, while the topics are quite specific: musicology, technical sciences, social work, medicine, dentistry, business, engineering, and humanities. The authors are working in mainly French university settings, with the two exceptions of Serbia and Italy, but represent a range of situations and teaching contexts. Readers will find, as Swales calls for, similarities and dissimilarities to their own situations and learn how others have dealt with problems they themselves might grapple with in their ESP work.

More recently, Ann Johns (2013), in introducing Paltridge and Starfield's *Handbook of English for Specific Purposes*, points out a related but somewhat different problem in reviewing the history of ESP research, that is,

---

1. Iowa State University, Ames, Iowa, United States of America; dandoug@iastate.edu

How to cite: Douglas, D. (2017). Foreword. In C. Sarré & S. Whyte (Eds), *New developments in ESP teaching and learning research* (pp. xv-xvii). Research-publishing.net. https://doi.org/10.14705/rpnet.2017.cssw2017.741

Foreword

> "making a clear distinction between research and practice. Unlike many other research areas in theoretical and applied linguistics, ESP has been, at its core, a practitioners' movement, devoted to establishing, through careful research, the needs and relevant discourse features for a targeted group of students" (p. 6).

This melding of research and practice, or perhaps, a symbiosis of research and practice, is another clear feature of the present volume. All of the authors are practicing, and practiced, ESP teachers who work in very different contexts, yet share an appreciation for essential characteristics of research in ESP: reviewing previous research and focussing on a gap in that research, constructing firm theoretical underpinnings for the study, understanding the dynamics of the context to be studied, taking into account the relationship between the context and the research methods, and drawing appropriate conclusions from the findings and relating them to the context.

In their introduction, the editors state that this volume intends to address key issues related to research in ESP teaching and learning by bringing together current research at the intersection of the theoretical and practical dimensions of English for Specific Purposes. I believe they have succeeded in this goal admirably. The chapters are well-written and well-edited, the topics are varied and interesting, the research techniques are similarly varied and carefully constructed to reflect the context, and the conclusions drawn are appropriately cautious and responsive to the ESP situation. The studies reported in this volume describe a number of research techniques that readers will find useful in their own contexts, including the use of questionnaires, action research, qualitative analyses, interviews, classroom observation, discourse analysis, and corpus development and use.

In this book, readers will discover a treasury of information they will find useful to their own understanding of research into ESP teaching and learning. No matter what area of ESP readers are working in, I believe they will be able to translate the research questions here into their own contexts. Such considerations as needs analysis and identification of learning objectives, the design of materials, learning tasks and assessment criteria, the measurement of student motivation in relation to course specificity, the use of audio-visual materials to encourage learning, approaches to the teaching of vocabulary, the use of narrative devices in the practice of scientific reading, the use of specific corpora in teaching, the use of fictional audio-visual material in the teaching of English for professional practice, and the use of pre-task activities to promote learning are all discussed.

In short, this volume encourages ESP practitioners to 'look across' ESP contexts for insights into their own situations. They will find, I believe, inspiration for research into classroom practice, learner attitudes and motivations, the benefits of subject specificity in relation to best practices of task design, and the use of various theoretical approaches to the teaching and learning of specific linguistic forms. Readers will find, too, I hope, encouragement to see themselves as professional ESP practitioners, engaging in research not simply to explore theoretical issues, but to identify the linguistic, intercultural, and personal needs of their learners which can be translated into learning materials and techniques. The chapters in this volume illustrate clearly the fundamental character of the specific purpose language enterprise: responding to differences in the beliefs, practices, and ways of speaking in the discourse communities learners wish to become members of. Much of what we do as ESP instructors is help our learners connect with the culture of their chosen fields. In order to do so, we need to understand the context of language use in those fields and devise methods to help learners integrate themselves communicatively into the specific professional, academic, and vocational culture. This book provides practical insights into that process.

References

Johns, A. (2013). The history of English for specific purposes research. In B. Paltridge & S. Starfield (Eds), *The handbook of English for specific purposes* (pp. 5-30). Wiley-Blackwell.

Swales, J. (1985). *Episodes in ESP*. Pergamon Press.

1. Introduction to new developments in ESP teaching and learning research

Shona Whyte[1] and Cédric Sarré[2]

In introducing the studies in this collective volume on research in teaching and learning English for Specific Purposes (ESP), we begin by considering the current context of language education in European universities, then examine key terms and concepts in our own vision of ESP didactics, before previewing the chapters selected for inclusion in the present volume. The book is divided into three sections, beginning with groundwork related to needs analysis and course design in disciplines as diverse as dentistry, musicology, and technical science, continuing with a closer look at particular ESP challenges related to lexico-grammar or genre, and in the final chapters moving on to innovative practice such as exploiting specialised corpora or television drama in the ESP classroom.

1. Background to ESP research

The present volume is one outcome of recent developments in a relatively new field of applied linguistics inquiry, at least as far as traditional European tertiary language education is concerned. In many such contexts, the main approach to language learning and teaching has historically involved cultural studies, particularly the literature, but also the social, economic, and political history of countries where the target language is spoken, generally referred to as Modern Foreign Language (MFL) studies. In the MFL tradition, research and teaching are closely intertwined, with literary scholars treating language and culture as an inseparable whole in both lecture theatres and scholarly journals.

---

1. Université Côte d'Azur, CNRS, BCL, Nice, France; shona.whyte@unice.fr

2. Université Paris-Sorbonne, ESPE de Paris, Paris, France; cedric.sarre@paris-sorbonne.fr

How to cite this chapter: Whyte, S., & Sarré, C. (2017). Introduction to new developments in ESP teaching and learning research. In C. Sarré & S. Whyte (Eds), *New developments in ESP teaching and learning research* (pp. 1-12). Research-publishing.net. https://doi.org/10.14705/rpnet.2017.cssw2017.742

Chapter 1

In this view, Language for Specific Purposes (LSP) is treated as a pedagogical concern, and one which can be met by changing teaching materials rather than teaching methods. The practical language needs of doctors, lawyers, or engineers, to name but these, are thus generally dealt with by instructors with MFL training, who replace literary texts relating to the target language culture with materials focussing on medical, legal, or engineering topics. More recently, some scholars involved in LSP teaching have also sought to pursue research here too, and the MFL background of these authors has naturally led them to focus on discourse analytic approaches to LSP. This is especially true in the case of ESP, defined as "a 'variety of English' that can be observed in a given perimeter of society, delineated by professional or disciplinary boundaries" (Saber, 2016, p. 2). Thus, text and discourse analysis have historically dominated ESP research (Hewings, 2002; Paltridge & Starfield, 2011), perhaps particularly in continental European work, where this domain-centered approach is made even more explicit in a concept termed "Specialised Varieties of English (SVE)" (Resche, 2015, p. 215).

In contrast, another tradition has developed with its roots firmly in teacher and learner needs in ESP. Interestingly, a good deal of the early work in notional-functional and communicative approaches to language teaching, which paved the way for today's Common European Framework of Reference for languages (CEFR), were motivated by ESP needs (Munby, 1978; Wilkins, 1972). Many have argued that this practical orientation, which characterises much applied research in ESP teaching and learning, has affected the academic standing of research in this "less glamorous" area (Hyland, 2006, p. 34), and ongoing tensions between practitioners and researchers remain a challenge. However, the time seems ripe to revisit the link between research, teaching, and learning in ESP contexts. In many European countries, we are witnessing a renewal of interest in teaching and learning of English which is tied to wider processes in the internationalisation of research and English as a global language, leading to more English Medium Instruction (EMI) and greater attention to English as a Lingua Franca (ELF). At the same time, increased use of technology in everyday and professional spheres is fuelling interest in Computer-Assisted Language Learning (CALL). Against this backdrop, a number of epistemological and methodological concerns in ESP have come to the fore (Sarré & Whyte, 2016).

2. Key terms and concepts in ESP didactics

Based on Petit's seminal work on theory-building in ESP leading to his 2002 representation, we have suggested the following definition of our field:

> "the branch of English language studies which concerns the language, discourse, and culture of English-language professional communities and specialised groups, as well as the learning and teaching of this object from a didactic perspective" (Sarré & Whyte, 2016, p. 150).

The multiple perspectives on ESP research mentioned in this definition – linguistic, cultural, discourse, and didactic – may suggest a "highly fragmented" field of research (Saber, 2016, p. 3). Yet these different aspects are complementary and can be viewed as dimensions of a "specific purpose language ability" (Douglas, 2001, p. 182), combining knowledge related to both language and content. Specific purpose language ability can be seen as a professional macro-skill comprising knowledge and competencies related to disciplinary, academic, or professional domains, and to particular modes of communication and relationships typical of each (Braud, Millot, Sarré, & Wozniak, 2017, pp. 37-38).

A key distinction in ESP teaching/learning research is between **pedagogy** and **didactics**. In previous work we have shown that, in continental Europe, "didactics is knowledge-oriented, a science which aims to understand how teaching leads to learning" (Sarré & Whyte, 2016, p. 142). This term is commonly used in research which is published in the national languages of mainland European countries, and which involves theorisation and distance from particular teaching contexts. Pedagogy, on the other hand, "is practice-oriented, concerned more with applied aspects of language teaching", best seen as "an applied component of didactics" (Sarré & Whyte, 2016, p. 142). This contrast is shown in Table 1.

Table 1. Didactics and pedagogy (from Sarré & Whyte, 2016)

Didactics	Pedagogy
• knowledge-oriented	• practice-oriented
• a distancing and theorising process	• a practical process
• main objective: the analysis of how teaching leads to learning	• main focus: teaching practices and education
• draws on various contributive sciences	• draws on didactic research = an applied component of didactics
• covers both SLA and foreign language education	• covers actors, curricula, content, context, and objectives

This distinction does not, however, hold in English-speaking research cultures, where only pedagogy is commonly used and the concept of didactics – covered within the overlapping fields of applied linguistics and Second Language Acquisition (SLA) research – is not, with important consequences for our field.

As "the general field of learning a non-primary language" (Gass, 1995, p. 3), SLA has provided the main theoretical foundation for language teaching and learning research in the English-speaking world since at least the early 1990's, both in the classroom (instructed SLA) and outside (naturalistic SLA). Indeed, the focus on learning in isolation from teaching allowed some researchers to sever all links with pedagogy as part of an endeavour to establish SLA as "an academic discipline in its own right" (Bygate, 2005, p. 568). In contrast, applied linguistics covers a broader interest in "language issues in any kind of real-world problem" (Bygate, 2005, p. 569) and involves "the theoretical and empirical investigation of real-world problems in which language is a central issue" (Brumfit, 1995, p. 27), or even more narrowly, "the pragmatically motivated study of language, where the term 'pragmatic' refers to the intention to address and not merely describe the real-world problems" (Bygate, 2005, p. 571). This principle of applicability links research to practice by viewing pedagogy as "an applied component of SLA" (Sarré & Whyte, 2016, p. 145), although the unilateral or unidirectional nature of this relationship has been contested. Arguing that SLA theory and other applied linguistics research has often failed to solve teaching problems, Widdowson (2017) suggests that:

> "in applied linguistics we need to reverse the dependency order of this relationship and analyse the problem first. [...] What disciplinary constructs and findings are of use can only be determined by analysing the problem first".

Instead of theory, or what he calls disciplinarity, being applied to real world issues, he claims real world issues should determine what type of disciplinarity is pertinent. Bygate, Skehan, and Swain (2001, p. 17) propose criteria to guide our choices concerning the focus, conceptualisation, and application of such research: it should meet the needs of language teachers, make sense to them, and produce results which they can use in their teaching. Here we have an agenda for an emerging academic discipline of ESP didactics drawing on:

- the numerous specificities of the field, identified across a multiplicity of ESP teaching and learning situations which call for a specific research framework (Sarré & Whyte, 2016);

- bilateral and bidirectional interactions between pedagogical practice and didactic theory (Sarré, 2017; Sarré & Whyte, 2016; Whyte, 2016); and

- a rich European tradition of language didactics research within this wider definition of applied linguistics.

The necessary link between didactic research and real-world contexts is perhaps one we more naturally preserve when publishing in our national languages than when writing in English, so it is helpful that this book should arise from a seminar at an English-language conference, the European Society for the Study of English (ESSE) in Galway (Milosevic, Molina, Sarré, & Whyte, 2016).

3. Current volume

Our goal in producing this volume was thus to attempt to bridge gaps between research and practice by offering strong research-based contributions in a wide range of ESP contexts. We have aimed to offer new theoretical and pedagogical insights for ESP practitioners and researchers alike, going beyond descriptions of ESP situations and/or programmes to involve sound research design and data analysis, anchored in previous ESP teaching and learning research. The nine papers in our collection cover a range of ESP domains: two in medicine (Birch-Bécaas & Hoskins; Franklin-Landi), two in technical science and engineering (Fries; Milosevic), two in social sciences (Johnson; Starkey-Perret, Belan, Lê Ngo, & Rialland), and three in humanities (Beaupoil-Hourdel, Josse, Kosmala, Masuga, & Morgenstern; Schug & Le Cor; Labetoulle). We present the studies in three subsections, beginning with needs analysis and course development, moving on to specific challenges in ESP teaching and learning, and concluding with some examples of innovative practice in our field.

3.1. Laying the groundwork: needs analysis, programme design, and course development

The papers in the first part of the collection address fundamental questions concerning the design of ESP activities and courses, such as learner needs and the design of courses and programmes. In contexts ranging from dental studies through to musicology and technical sciences, the studies report on action research undertaken to improve the quality of English courses offered to students in their various institutions. In each case, the authors are ESP practitioners confronted with particular difficulties and challenges which they have sought to address with reference to research in SLA, educational theory, and ESP research itself. The three papers highlight the importance of needs analysis, as well as the problems of trying to both identify and meet the multiple and often conflicting requirements of students, language instructors, lecturers in content areas, as well as institutions.

Milosevic designed a small pilot test of audiovisual resources for teaching English for technical sciences in Serbia, using an experimental/control design and measuring reading comprehension and translation of key terms. Her students preferred and performed at least as well with teaching materials using video rather than text alone, and the study seems to justify wider trials with more teaching units and more controlled testing. **Labetoulle**'s research also started from needs analysis, particularly the challenges of a large, heterogeneous student population and a heavy workload for language instructors in musicology at a large French university. In her study, the adaptation of a Complex Dynamic Systems (CDS) framework from the perspective of didactics ergonomics (Bertin, Gravé, & Narcy-Combes, 2010; Rivens-Mompean, 2013) allowed practitioners to take a number of factors into account in the design and implementation of new ESP courses, though the chapter suggests it may be difficult if not impossible to satisfy all needs involved. In **Birch-Bécaas and Hoskins'** study of a final year dentistry course, the main focus is on the conception of an ESP task which met a number of demands from students, language instructors, and dentistry lecturers, as well as institutional assessment criteria. The study reports high levels of participant satisfaction achieved by integrating needs analysis, second language research, and both teacher and learner perspectives.

3.2. Building confidence: addressing particular difficulties

The second part of the collection takes a closer look at specific problems occurring in ESP teaching. **Schug and Le Cor** follow Waninge (2015) in a CDS approach to the complex question of learner motivation, tracking four individual students in four different ESP and non-ESP classes at different time scales (i.e. over three-hour class periods, and over three-month courses). Their findings suggest a great deal of individual variation, not all related to ESP. **Fries** reports on a pedagogical intervention concerning a particular aspect of lexico-grammar: compound nouns in engineering discourse, drawing on cognitive semantics (Langacker, 1987). Her analysis of learners' use of the target structure in a writing assignment and on presentation slides suggests little quantitative but a possible qualitative advantage for her experimental approach to teaching compound nouns via cognitive semantics. **Beaupoil-Hourdel and colleagues** addressed the problem of supporting humanities students in reading scientific articles, drawing on classic work in script theory (Schank & Abelson, 1977) and more recent efforts to improve scientific communication using narrative techniques (Luna, 2013; Olson, 2015). The paper describes the creation of teaching materials based on devices such as narrative elements and the dramatic arc, and tests their efficacy with measures of learner performance and attitude. These studies include close analysis of relatively large amounts of data, be these compound nouns in learner writing, classroom observations of learner engagement, or measures of reading proficiency, all with the aim of investigating the effect of ESP instruction.

3.3. Moving ahead: towards new practices

In the third and final section of our collection we look at new practices involving different approaches to materials design and pedagogical support for ESP learning. **Johnson** situates her study of ESP for social work at an Italian university in a wider context which shows how both the Bologna process and the more recent migrant crisis have had far-reaching effects. She advocates a holistic approach to ESP, drawing on corpus tools to constitute and exploit a specific monitor corpus for these students, using a range of tools to identify and raise awareness of patterns of native-speaker usage. **Franklin-Landi** follows a recent tradition in French higher education to exploit fictional representations of specific domains, such as medicine or law in ESP teaching. She charts student perspectives on the medical TV series *Grey's Anatomy* following classroom activities based on a video extract, highlighting advantages and risks in using such material as a pedagogical resource. The final paper in this section by **Starkey-Perret and colleagues** also examines the effects of a particular pedagogical intervention, this time drawing on Schmidt's (1990) notion of noticing to investigate the impact of focus on form activities prior to the main task. Students in this study were enrolled in a business English course and the authors compared frequency and accuracy of lexico-grammatical features in the production of students who chose to complete the pre-task activities with those who did not, as part of a wider research programme into the effectiveness of task-based language teaching with this population. The study was complicated by the high drop-out rate common in certain French undergraduate programmes, as well as the wide range of features included in the intervention. This third section of the volume reminds readers of the sheer variety and complexity of ESP contexts and the correspondingly broad spectrum of dimensions in need of attention, from needs analysis to pedagogical resources and teaching activities.

4. Conclusion

In our view, this collection of studies raises questions of relevance to the field of ESP teaching and learning research with reference to three main areas: (1) the balance of content and language aspects of ESP teaching; (2) factors related to ESP learners and second language learning; and (3) issues of research design and methodology. The debate about how best to coordinate the development of disciplinary knowledge and the linguistic means to express it is far from over, and several language practitioners in this volume offer suggestions in this respect. Birch-Bécaas and Hoskins note their difficulties as language instructors in drawing students' attention away from disciplinary information toward linguistic form, as well as their students' appreciation of content instructors' efforts to 'play the game' and use English to help create a natural context for language practice. Franklin-Landi voices a common concern among language instructors regarding their own legitimacy in a domain where they cannot claim content expertise, and sees specialised fiction as a kind of third space where errors committed by non-specialist writers and actors provide both motivation and resources for fruitful discussion in the target language.

Second, a number of chapters in this book address issues related to learners and language learning which are not specific to ESP. Many of the studies investigate attitudinal questions related to motivation and stress in the language classroom, and several also touch on problems of poor attendance and high drop-out rates. The complexity of these issues has led some to CDS as a suitable framework for accommodating a range of variables in a systematic manner (Labetoulle; Schug & Le Cor). Others have sought to gauge learner views of particular teaching and learning activities via questionnaires and interviews (Beaupoil-Hourdel et al.; Birch-Bécaas & Hoskins; Franklin-Landi; Milosevic) to support students in progressive approaches to new competencies, such as critical reading of research (Beaupoil-Hourdel et al., Birch-Bécaas & Hoskins), awareness-raising of specificities of genre (Johnson), or simply to anticipate absenteeism in research design (Starkey-Perret et al.).

Finally, the studies in this volume have highlighted a number of challenges inherent in classroom research at this level. A number of chapters reported on research designs based on experimental and control groups and providing quantitative analysis of student performance (Milosevic; Fries; Starkey-Perret et al.). In each case it proved difficult to establish clear-cut statistically significant intergroup differences, once again highlighting the complex teaching and learning situations experienced by many ESP practitioner-researchers. We might expect the way forward to lie with interdisciplinary teams combining linguistic and statistical experience, controlling for more variables and perhaps also sharing instruments. It is no doubt important, too, to measure not only language accuracy but also complexity and fluency as essential components of communicative competence and indicators of interlanguage development.

In conclusion, the chapters in this collection remind us of the inseparable nature of pedagogy and didactics. While 'researched pedagogy' (Bygate, 2005; Bygate et al., 2001) might appear at first sight a contradiction in terms, we hope we have made the argument for both more research-informed practice and more practice-driven research. Research in ESP teaching and learning has often been criticised for a lack of theoretical underpinning, making findings difficult to generalise to new contexts (Sarré, 2017), or for lack of applicability to actual language teaching (Master, 2005; Widdowson, 2017). The present volume attempts to address these criticisms by building on previous research, reporting on a variety of contexts, and representing a range of theoretical frameworks and methods. Results are reported with an emphasis on applicability in order to strengthen links between didactics and pedagogy, and to suggest future directions likely to benefit practitioners and researchers alike. In this way, despite the inescapably specific contexts of our studies, and other limitations on their generalisability, this collection encapsulates current trends and new developments in ESP teaching and learning research in Europe and we hope makes a small but valuable contribution to the field of ESP didactics.

References

Bertin, J.-C., Gravé, P., & Narcy-Combes, J.-P. (2010). *Second language distance learning and teaching: theoretical perspectives and didactic ergonomics: theoretical perspectives and didactic ergonomics*. IGI Global. https://doi.org/10.4018/978-1-61520-707-7

Braud, V., Millot, P., Sarré, C., & Wozniak, S. (2017). Quelles conceptions de la maîtrise de l'anglais en contexte professionnel ? Vers une définition de la compétence en anglais de spécialité. *Mélanges CRAPEL, 37*, 13-44.

Brumfit, C. (1995). Teacher professionalism and research. In G. Cook & B. Seidelhofer (Eds), *Principle and practice in applied linguistics* (pp. 27-41). Oxford University Press.

Bygate, M. (2005). Applied linguistics: a pragmatic discipline, a generic discipline? *Applied Linguistics, 26*(4), 568-581.

Bygate, M., Skehan, P., & Swain, M. (2001). *Researching pedagogic tasks – second language learning, teaching and testing.* Longman.

Douglas, D. (2001). Three problems in testing language for specific purposes: authenticity, specificity and inseparability. In C. Elder, A. Brown, E. Grove, K. Hill, N. Iwashita, T. Lumley, T. McNamara & K. O'Loughlin (Eds), *Experimenting with uncertainty: essays in Honour of Alan Davies* (pp. 45-52). Cambridge University Press.

Gass, S. (1995). Learning and teaching: the necessary intersection. In F. Eckman et al. (Eds), *Second language acquisition theory and pedagogy* (pp. 3-20). Erlbaum.

Hewings, M. (2002). A history of ESP through English for specific purposes. *English for Specific Purposes World, 3*(1).

Hyland, K. (2006). The 'other' English: thoughts on EAP and academic writing. *The European English Messenger, 15*(2), 34-38.

Langacker, R. W. (1987). *Foundations of cognitive grammar: theoretical prerequisites.* Stanford University Press.

Luna, R. E. (2013). *The art of scientific storytelling.* Amado International.

Master, P. (2005). Research in English for specific purposes. In E. Hinkel (Ed.), *Handbook of research in second language teaching and learning.* Lawrence Erlbaum Associates.

Milosevic, D., Molino, A., Sarré, C., & Whyte, S. (2016). *Teaching practices in ESP today.* Seminar 14, ESSE Conference (European Society for the Study of English), Galway, Ireland.

Munby, J. (1978). *Communicative syllabus design: a sociolinguistic model for designing the content of purpose-specific language programmes.* Cambridge University Press.

Olson, R. (2015). *Houston, we have a narrative: why science needs story.* University of Chicago Press. https://doi.org/10.7208/chicago/9780226270982.001.0001

Paltridge, B., & Starfield, S. (2011). Research in English for Specific Purposes. In E. Hinkel (Ed.), *Handbook of research in second language teaching and learning, volume 2* (pp. 106-121). Routledge.

Petit, M. (2002). Éditorial. *ASp*, 35/36, 1-3.

Resche, C. (2015). Mapping out research paths to specialised domains and discourse: the example of business cycles and financial crises. *Journal of Teaching English for Specific and Academic Purposes*, *3*(2), 215-228.

Rivens-Mompean, A. (2013). *Le Centre de Ressources en Langues: vers la modélisation du dispositif d'apprentissage*. Presses Univ. Septentrion. https://doi.org/10.4000/books.septentrion.16720

Saber, A. (2016). Éditorial: Immanuel Kant and ESP's new frontier. *ASp*, *69*, 1-6.

Sarré, C. (2017). La didactique des langues de spécialité : un champ disciplinaire singulier ? *Les Langues Modernes, 3*, 53-64.

Sarré, C., & Whyte, S. (2016). Research in ESP teaching and learning in French higher education: developing the construct of ESP didactics. *ASp, 69*, 139-1164. https://doi.org/10.4000/asp.4834

Schank, R. C., & Abelson, R. (1977). *Scripts, goals, plans, and understanding*. Erlbaum.

Schmidt, R. (1990). The role of consciousness in second language learning. *Applied Linguistics, 11*(2), 129-158.

Waninge, F. (2015). Motivation, emotion and cognition. Attractor states in the classroom. In Z. Dörnyei, A. Henry & P. D. MacIntyre (Eds), *Motivational dynamics in language learning* (pp. 195-213). Multilingual Matters.

Whyte, S. (2016). Who are the specialists? Teaching and learning specialised language in French educational contexts. *Recherches et pratiques pédagogiques en langue de spécialité, 35*(3). https://apliut.revues.org/5487

Widdowson, H. (2017). *Disciplinarity and disparity in applied linguistics*. Plenary address, British Association for Applied Linguistics, Leeds, September.

Wilkins, D. A. (1972). An investigation into the linguistic and situational content of the common core in a unit/credit system. *Systems development in adult language learning*. Council of Europe.

Section 1.

Laying the groundwork: needs analysis, programme design, and course development

Using video materials in English for technical sciences: a case study

Danica Milosevic[1]

Abstract

In the digital era, university instructors working in English for Technical Sciences (ETS) have opportunities, some might say obligations, to use audio-visual resources to motivate students. Such materials also call on cognitive and constructivist mechanisms thought to improve uptake of the target language (Tarnopolsky, 2012). This chapter reports on a small-scale study on the effectiveness of audio-visual materials in the development of comprehension and vocabulary skills in ETS. The study was conducted at the College of Applied Technical Sciences in Serbia and involved three groups of students: an audio-visual group exposed only to the video resources, a control group using web print-outs, and a combined group with access to both types of materials. Student performance was compared, and evaluation questionnaires developed to collect feedback from all participants. This chapter details the entire process of the study, elaborates on the findings, and gives an analysis of the data obtained to advocate for a more extensive use of audio-visual material in ETS practice.

Keywords: audio-visuals, comprehension skills, vocabulary skills, t-test, English for specific purposes, ESP.

---

1. Independent researcher, Nis, Serbia; danicamil@yahoo.com

How to cite this chapter: Milosevic, D. (2017). Using video materials in English for technical sciences: a case study. In C. Sarré & S. Whyte (Eds), *New developments in ESP teaching and learning research* (pp. 15-30). Research-publishing.net. https://doi.org/10.14705/rpnet.2017.cssw2017.743

Chapter 2

1. Introduction

The world we live in today is constantly changing due to technological discoveries that are shaping our daily routine and becoming an inextricable part of our daily lives. These changes are so rapid that sometimes it is not possible to notice them in time and react appropriately. English for Specific Purposes (ESP) practitioners who teach Technical English (TE) courses, however, seem to be on a ceaseless quest for innovations in domains of science and technology in order to satisfy the needs of their students. They need to use the available material on high-tech achievements in their ESP classroom, and also need to modernise their ESP courses, thus providing students with substantial professional language input and updated information.

YouTube, which offers a wide variety of video clips that deal with professional topics, is one of the most useful and effective teaching aids in the modern TE environment. It contains all the latest professional information to be found in documentaries, popular science shows, tutorials, lectures, advertisements, and much more. If properly selected, these materials can operate as audio-visual resources that bring real life into the classroom, present students with every-day professional situations, and play the role of valuable authentic material, which altogether is of great importance in the constructivist system of ESP teaching. That is why audio-visual instruction material in the form of YouTube clips is considered a tool that can modernise the teaching/learning process and meet the interests and needs of students who crave to be in touch with technology. The study to be presented in this chapter discloses how students of technical sciences react to video resources in ESP classes and reveals the achievements of students instructed by such teaching aids.

2. Recent research on the audio-visual resources in EFL/ESP courses

Studies that have been conducted recently on the use of Audio-Visual Resources (AVRs) in English language teaching and learning mainly explore the attitudes of general English language teachers and students towards audio-visual resources as teaching tools. For example, such studies investigate how AVRs affect primary school children (Parvin & Salam, 2015) and secondary school children in learning English (Ode, 2014), or how they influence the performances of university students (Kausar, 2013; Mathew & Aldimat, 2013) in their English language and literature classes. They also give insight into how secondary school English literature teachers perceive the role of AVRs in motivating students to read literature and develop their critical and creative skills (Yunus, Salehi, & John, 2013), or, for instance, show the extent to which English Language (EL) teachers are satisfied with the quality of the audio-visual material provided by libraries (Ashaver & Igyuve, 2013). All of these studies were mainly conducted by means of evaluation questionnaires that provide both qualitative and quantitative data highly supportive of the use of audio-visual resources in the EFL classroom.

In addition, there are studies that deal with the impact of AVRs on particular language skills. Such studies (1) focus on the usefulness of AVRs in building students' competence in pronunciation (Gilakjani, 2011; Ranasinghe & Leisher, 2009), (2) elaborate on the effects of AVRs in promoting target language communication, that is, the speaking skills of students (Çakir, 2006; Natoli, 2011; Ramin, Reza, & Nazli, 2014), or (3) study the importance of video instruction in teaching vocabulary (Gross, 1993; Wright & Haleem, 1991). They, furthermore, take listening skills into consideration to investigate how visual material can be used as a teaching strategy for improving listening comprehension skills (Potosi, Loaiza, & Garcia, 2009), showing that video sessions contribute to a better mastery of listening skills and a greater motivation for students to participate in conversation. Studies devoted to language skills also test how audio-visual, audio, and video tools affect the writing skills of EL students (Ghaedsharafi & Bagheri, 2012), finding out that most advanced skills in writing are displayed by the students in the audio-visual group, whereas the weakest skills are seen among students in the video group. The advantages of audio-visual resources over other teaching tools, as well as their classifications and their specific features, are explored by other authors as well (Asokhia, 2009; Daniel, 2013; McNaught, 2007; Viswanath & Maheswara, 2016).

Chapter 2

Recent research on audio-visual resources in ESP environments is less common. A significant study was conducted by Al Khayyat (2016), who determined that ESP skills were better developed in students who had been instructed by a combination of audio-visual resources and computerised material than in students whose instruction was based on conventional materials. This conclusion was made by comparing the students' speaking and writing abilities. As far as the specific terminology among ESP students is concerned, a study was carried out with the intention of determining how audio-visual resources could be used to develop students' vocabulary skills, confirming that AVRs can positively impact not only vocabulary but also writing skills in ESP students (Lin, 2004).

In addition, perceptions of ESP teachers regarding the use of AVRs in teaching were also explored. One such study was conducted by Şahin and Şule Seçer (2016) among the English teachers in an aviation high school to see how efficiently they had been using video materials as warm-up activities in their classes. Focus group interviews used by the authors revealed that AVRs were not as frequently incorporated in classes as expected. The biggest obstacles for implementation of AVRs seemed to be the shortage of time to use video material, the inadequacy of the equipment used, or insufficient computer literacy of teachers who needed to cope with technical problems that occurred in class.

Mutar (2009) is one of the rare authors who studied the impact of AVRs on a technical English course. He compared the overall scores of students who were using audio-visual materials with a control group with no access to such materials. His study revealed a significant difference between the two samples, with the experimental group scoring significantly higher. However, this finding should not be taken for granted due to the fact that Mutar's evaluation instrument had certain limitations concerning the scope of exercises for testing comprehension and vocabulary skills.

The study to be presented in this chapter, however, was not conducted to explore the cumulative effect of AVRs on the teaching/learning process in one ESP course, but rather to check the effects after a single use of video resources in class. Since there is a possibility that ESP practitioners might be unable to use video material during each and every class throughout the course, due to the obstacles previously mentioned by Şahin and Şule Seçer (2016), the sporadic usage of video resources is something which is more realistic and therefore should be tested as such. In addition, this study focussed only on comprehension and vocabulary skills to explore more deeply the impact that AVRs could have on these particular linguistic competencies. The main purpose was to see if the ETS material at the tertiary level of education was easier to grasp and memorise through the application of video resources; thus the effect of AVRs on cognitive capacities was also of interest in this study.

3. Benefits of implementing audio-visual resources in ESP classes

Speaking of the effectiveness of video materials in teaching, Ode (2014) stated that "audiovisual resources do not only increase the motivation of teachers and learners, [but] they [also] add clarity to the topic taught and make learning more interesting" (p. 195). Ode is just one of the authors who claims that video materials bring added value, making classes more appealing and making teaching material easier to understand. Mannan (2005) and Dike (1989) also spoke in favour of AVRs as tools for clarifications on complex subject-matter. Mannan (2005) stated that AVRs "help the teacher to clarify, establish, correlate and coordinate accurate concepts, interpretations and appreciations, and enable him to make learning more concrete, effective, interesting, inspirational, meaningful and vivid" (p. 108). Unlike professional texts which can be full of ambiguity, abstract words, and meanings, audio-visual materials can present concrete examples and eliminate abstraction from language. This is achieved through a variety of linguistic and non-linguistic cues that are displayed in a video. This way, abstract ideas which are accompanied by concrete visual presentations can become more understandable to the audience. When watching the material and listening to it simultaneously, students can almost immediately test their understanding of a certain video content, their audio and visual capacities being stimulated at the same time. They can rely on many paralinguistic features too, like mimicking, gestures, postures, or attitudes that can assist them in grasping the meaning of the study material, which can facilitate learning of their ESP content.

Apart from these advantages, studies have shown that video instruction also has beneficial effects on retention of the material which is being taught (Barry, 2001; Clark & Lyons, 2004; Njoku, 1980; Paivio & Clark, 1991). For instance, Njoku (1980) pointed out that AVRs promote learning and make it more durable due to the fact that they have an impact on more than one sense at a time. Since audio-visual materials engage our visual and auditory capacities simultaneously, they lead to the creation of strong conceptual images in our minds. These conceptual images are associations of pictures with words, which once installed in our minds are more likely to stay there on a permanent basis and be recalled more easily.

Apart from all this, AVRs can expose students to authentic language which is accurate, vivid, conversational, and related to real-life experience. This kind of language input can have a positive effect on intensifying the use of the target language in ESP classes, as Wilson (2001) implies by saying that "the use of visuals enhances language learning on one hand and increases the use of target language on the other" (p. 11). In her opinion, when input language is reused in creative ways, it can help students to improve their communication, which will become more proficient over time. This opinion is shared by Tarnopolsky (2012) who advocates the use of audio-visual material as authentic input in ESP. In his constructivist blended learning approach, which combines traditional methods with modern web-based technologies, Tarnopolsky stresses the importance of practical experience in e-learning. As a supporter of the input-output theory, which requires that students are given a chance to practice what they have learnt, he proposes an entire set of ESP activities (role-plays, case studies, simulations, etc.) with recommendations for integration of the four skills. In this process, he claims that post-viewing activities are as important as the viewing activity itself: "only the online resources which blended learning activates in the learning process may provide sufficient authentic materials for modeling professional activities and professional target language communication in the classroom" (Tarnopolsky, 2012, p. 15).

This small-scale study relies on Tarnopolsky's blended learning approach, which states that "successful knowledge and skills *constructing* is achieved owing to students' regular Internet research on professional sites in English" (Tarnopolsky, 2012, p. 15, emphasis in original). Therefore, in this small-scale study, internet resources are implemented as instructional materials, and post-viewing activities are employed with the intention of testing whether students with video input learn better than the students who are not exposed to video materials. This study was part of an initial needs analysis to determine curricular priorities among applied technical science students at the Serbian university of Nis for the coming academic year. Since many students who enroll at the College of Applied Technical Sciences have never had ESP classes during their high school education but have rather attended classes of general English, it was significant to see how they, as first year students, react to ESP and whether they preferred traditional teaching based on written materials to the use of video resources. Perhaps, more importantly, the study also sought to determine any differences in achievement across instructional conditions.

4. A small-scale study: effects of AVRs on comprehension skills and the use of vocabulary in TE settings

As already noted, the goal of this study was to determine to which extent AVRs have positive effects on student comprehension and vocabulary skills in the field of ESP after just a single exposure to video materials. The experiment was carried out at the beginning of the ESP course with 70 students in three intact classes of first year students (aged 19-20) of Modern Computer Technologies, Communication Technologies, and Industrial Engineering departments, who respectively formed the video, text/image, and feedback groups.

The first two groups were taught an ESP unit lasting for one 45-minute session on a topic in mobile technology and Active-Matrix Organic Light-Emitting Diode (AMOLED) displays. In each session, the students first completed a five-minute pre-test, were given a ten-minute mini-lesson on AMOLED displays, then completed a 25-minute test, before a five-minute feedback questionnaire. The mini-lesson format differed across groups: the video group (27 students) was exposed to audio-visual material, while the text/image group (26 students) was given only written materials. A third group (17 students) was asked to rate both types of pedagogical resources only in a feedback session after a 20-minute mini-lesson. The null hypothesis was: There is no significant difference between the test results of the students who were exposed to the video and those who were not. The results are reported in three parts: pre-test, main test and feedback (questionnaire).

4.1. Prior knowledge

Before the main testing, all 53 students from the video and text/image groups took a pre-test, the purpose of which was to check the students' prior knowledge of AMOLED displays used in smart mobile devices. The pre-test encompassed two simple questions: (1) Have you ever heard of AMOLED displays? and (2) What do you know about AMOLED technology?, to which the students provided written answers. The pre-test allowed us to rule out inter-group differences in background knowledge as well as a potential ceiling effect (where prior knowledge is so well developed as to render further instruction superfluous). Only 25 students (13 video, 12 text/image) had heard about AMOLED and were able to provide only the basic information. Therefore, the pre-test disclosed that the topic could be classified as a new instruction material in vocational English, and that there were no significant differences in prior knowledge across groups, as can be concluded based on the results obtained, shown in Table 1.

Table 1. Pre-test results

Average score (total = 3)	Video group N =27	Text group N =26
0-1.0	21	22
1.5-3	6	4
Mean	0.78	0.67
SD	1.01	0.93

4.2. Pedagogical materials and instruments

The second phase consisted of the instruction part and the assessment part. During the instruction section, the video group was exposed to the video material about AMOLED displays ("Why AMOLED enables the best displays for smartphones"[2]). This video is a short Samsung advertisement recorded in the form of a cartoon. The advertisement presents AMOLED displays in a very good manner, drawing attention to all the important technical details of this new piece of technology. It features a young female character with whom the students could easily identify with as mobile phone users. The video is especially useful because it contains a segment in which all the major performances of AMOLED displays are summed up, which makes the material easier to memorise.

The text/image group was given two texts to read about AMOLED displays, from 2 different Internet sources[3]. By reading both texts, the students could approach the topic from two different angles and memorise the reading material better. The teaching materials provided both groups with similar learning opportunities: both the video and the texts were based on authentic, authoritative content. Whereas the video contained additional information on AMOLED features, the text contained more details on the structure of AMOLED displays. However, the test that was administered to both groups did not address these differences and focussed only on the points shared by both the text and the video.

4.3. Comprehension and vocabulary tests

Following the ten-minute session with one type of resource or the other, each group was immediately tested on their comprehension and vocabulary skills via a written test. The test consisted of two exercises: the first one had four comprehension questions about the topic, each graded from 0 to 3 on the basis of meaning (vocabulary) rather than grammatical accuracy or spelling. The vocabulary exercise comprised of a list of 13 words/phrases to be translated

---

2. https://www.youtube.com/watch?v=EMYksNkH868

3. See supplement: https://research-publishing.box.com/s/03dgtaxps5terq8ckev0ipmvggkfa9o1

Chapter 2

from Serbian into English, including transparent and technical terms, which were graded on accuracy and spelling.

5. Results

The results of the main test are shown in Table 2.

Table 2.   Main test results

Average score (total=18.5)	Video group N=27	Text group N=26
0-9.4	18	25
9.4-18.5	8	2
Mean	5.73	5.29
SD	1.88	1.14

A t-test performed on these group means is significant at $p<0.1$ (not the more commonly accepted p 0.05 level), suggesting a marginal advantage for the video group. This finding is supported by measures of average length of responses: the video group wrote longer, more elaborate answers to the questions in the comprehension exercise, as shown in Table 3.

Table 3.   Comprehension exercise results

Item	Video group (average number of words per answer)	Text/image group (average number of words per answer)
1.	2.74	1.5
2.	1.37	2.23
3.	1.56	1.08
4.	2.89	2.12

In order to rank student scores, the assessment scale shown in Table 4 was established. While eight students from the video group obtained a grade of about 9.4, set as the passing grade, only two students from the text/image group did so. These two students were the outliers whose performance brought up the overall group average.

After the test, both groups completed a customised questionnaire where they gave their opinion on ten statements (using a Likert scale) and wrote a comment in the box designed for an additional open-ended question. The questionnaire analysis determined the most frequently selected answer for each question in each group separately, as it can be seen in Table 5.

Table 4. Assessment scale

Grade	No. of points	Video group (N=27)	Text group (N=26)
5	< 9.4	19	24
6	9.4 -11.22	3	0
7	11.23-13.05	2	0
8	13.06-14.88	1	0
9	14.89-16.71	1	2
10	16.72-18.5	1	0

Table 5. Most frequently selected answers

No	Questions	Video feedback average (27 students)	Text/image feedback average (26 students)
1	The video/text was too long.	2.8	3.1
2	The video/text was boring.	2.8	2.9
3	I have understood the given material very well.	3.7	3.6
4	There were many technical words that were too difficult to understand.	2.3	2.4
5	I have learned many new words.	2.4	2.7
6	It was difficult for me to answer the questions about AMOLED displays.	2.8	2.7
7	It was difficult to translate the terms from Serbian into English.	2.4	2.3
8	I liked the material we were doing in class.	2.9	3.2
9	I liked the activities I was asked to do.	2.6	3.0
10	These exercises were useful.	3.2	3.4

Individuals in the video group assessed the video material as reasonably long and motivating enough. They stated that they could understand the video very well and that it was neither easy nor hard to answer the comprehension questions. Their opinion in general was that the technical words could be understood and translated without any particular difficulties. Also, they mentioned that many of the technical terms were already familiar to them. This could be the only negative comment, but justification for this kind of answer can be found in the clear and concrete language presented in the video supported by the visual stimuli that contributed to a better understanding of the study material. Also, the students said that they were not overwhelmed with the activities they were asked to do in class, but they did not say that the material and the exercises were boring or useless. Most students gave a comment in the personal opinion box that the language experiment was interesting, and only one student protested against advertisement use in class due to its influence on the opinion of students as buyers of this kind of equipment.

Individuals in the control group, on their part, thought that the text was of moderate length and stimulating enough. They said that they could understand the text and answer the questions with no particular difficulties. They encountered no problems when it came to translating the necessary terms. Again, the students had the impression that there were not many technical words that were unfamiliar to them. The conclusion is that they may have encountered this vocabulary somewhere else, in some other context. However, the material and the activities done in class turned out to be appealing to this group. Most students thought that the exercises were useful, so the language experiment was agreeable to this group as well, which leads to the overall conclusion of this survey that both groups of participants were satisfied with the given material and the activities they were doing in class.

In addition, the analysis of the questionnaires for the video and text/image groups showed that the students had very high opinions of themselves and their performances on the main test, which were not borne out by analysis of the results reported earlier. In reality, it turned out that the students did not have as many problems when it came to translating phrases from Serbian into English as they had with the written part of the test which required them to provide technical explanations of AMOLED, its structure, and functions. Maybe the students were familiar with the vocabulary, but they could not apply that vocabulary with a concrete purpose to elaborate on the topic. This is probably where the reason lies for the discrepancy between the students' actual scores and their personal opinion on their test efficiency.

For further insight into the two types of teaching materials, a group of 17 students from the Industrial Engineering department were also asked to give their personal opinion after reading two professional texts on AMOLED and then watching the video during their ESP class. The idea was to allow them to express their opinions and make comparisons between the two types of resources: the video and the texts. They were asked to write down their impressions on a piece of paper, stating which mode of instruction they preferred and why.

Almost all of their comments were in favour of the video, implying that: (1) it is not as boring as the text; (2) it contains more information; (3) it explains the subject-matter in a better way; (4) it is more effective since it consists of a picture, text, and sound; (5) it is easier to grasp; and (6) it allows for better retention of language elements. One student also mentioned that he/she could easily identify with the video. The analysis of this evaluation has thus shown a great response of the students towards video materials, which are seen as very motivating and useful tools for studying the target professional language.

6. Conclusion

This paper has reviewed research on the role of video materials in facilitating ESP learning by improving the retention of facts, making classes more appealing, and contributing to the ESP environment by promoting and using authentic language. All of these features of AVRs are possible due to the fact that AVRs "create a concrete basis for conceptual thinking" (Ode, 2014, p. 198), and "make abstract ideas more concrete to learners" (Ode, 2014, p. 195) while also allowing learners "to develop a holistic understanding that the words cannot convey" (Ramírez,

2012, pp. 20-21), encouraging students to process language information more easily and participate in authentic communication and production.

The paper reported on a small case study carried out as part of an initial needs analysis for the first year ESP classes in technical sciences programmes. A comparison of instructional materials based on video or text has shown that even after a single use in class, video resources appear to be more popular and marginally more effective than text-based teaching materials. Since this was only a small-scale study, additional research is of course necessary to support these tentative findings that students exposed to AVRs may achieve better results in comprehension exercises than the students who were instructed by traditional methods and may have an advantage in both comprehension and language production. The study confirms previous research suggesting that students find video materials a much more pleasing and effective type of learning resource than exclusively written resources. Further research involving more students, longer interventions, and a variety of pedagogical materials and evaluation methods would allow stronger claims about instructional effects in relation to pedagogical resources in this area of ESP teaching and thus offer more insights into the effectiveness of video resources in technical science contexts.

References

Al Khayyat, A. (2016). The impact of audio-visual aids (AVA) and computerized materials (CM) on university ESP students' progress in English language. *International Journal of Education and Research, 4*(1), 273-282.

Ashaver, D., & Igyuve, S. M. (2013). The use of audio-visual materials in the teaching and learning processes in colleges of education in Benu state-Nigeria. *Journal of Research & Method in Education, 1*(6), 44-55, 2320-7388.

Asokhia, M. O. (2009). Improvisation/teaching aids: aid to effective teaching of English language. *International Journal of Education and Science, 1*(2), 79-85.

Barry, A. M. (2001). Faster than the speed of thought: vision, perceptual learning, and the pace of cognitive reflection. *Journal of Visual Literacy, 21*(2), 107-122. https://doi.org/10.1080/23796529.2001.11674574

Çakir, I. (2006). The use of video as an audio-visual material in foreign language teaching classroom. *The Turkish Online Journal of Educational Technology,* 5(4), 67-72. http://www.tojet.net/articles/v5i4/549.pdf

Clark, R. C., & Lyons, C. (2004). Graphics for learning: proven guidelines for planning, designing, and evaluation visuals in training materials. Pfieffer. https://doi.org/10.1002/pfi.4140431011

Daniel, J. (2013). Audio-visual aids in teaching English. *International Journal of Innovative Research in Science, Engineering and Technology,* 2(8), 3811-3814.

Dike, H. L. (1989). *Strategies for producing instructional materials.* The Government Printers.

Ghaedsharafi, M., & Bagheri, M. S. (2012). Effects of audiovisual, audio and visual presentations on EFL learners' writing skill. *International Journal of English Linguistics,* 2(2), 113-121. https://doi.org/10.5539/ijel.v2n2p113

Gilakjani, A. B. (2011). A study on the situation of pronunciation instruction in ESL/EFL classrooms. *Journal of Studies in Education,* 1(1), 1-15.

Gross, B. D. (1993). Tools for teaching. Jossey Bass Publishers.

Kausar, G. (2013). Role of teachers' and students' beliefs in English language learning at federal colleges of Pakistan. PhD thesis.

Lin, L. F. (2004). EFL learners' identical vocabulary acquisition in the video-based CALL program. *Asian EFL Journal,* 12, 37-49.

Mannan, A. (2005). *Modern education: audio-visual aids.* Anmol Publications.

Mathew, N., & Aldimat, A. (2013). A study on the usefulness of audio-visual aids in EFL classroom: implications for effective instruction. *International Journal of Higher Education,* 2(2), 86-92. https://doi.org/10.5430/ijhe.v2n2p86

McNaught, A. (2007). Moving images and sound: inclusive and accessible. In C. Grant and I. Mekere (Eds), *Moving images knowledge and access* (pp. 29-33). British Universities Film and Video Council.

Mutar, S. S. (2009). The effect of using technical audio visual-aids on learning technical English language at technical institutes. *Misan Journal for Academic Studies,* 8(15), 1-12.

Natoli, C. (2011). *The importance of audio-visual materials in teaching and learning.* https:www.helium.com/channels/224-early-childhood-ed

Njoku, P. A. (1980). *Practical hints on principles of education.* Africa Educational.

Ode, E. O. (2014). Impact of audio-visual (AVS) resources on teaching and learning in some selected private secondary schools in Makurdi. *IMPACT: International Journal of Research in Humanities, Arts and Literature,* 2(5), 195-202. http://oaji.net/articles/2014/488-1404472770.pdf

Paivio, A., & Clark, J. M. (1991). Dual coding theory and education. *Educational Psychology Review, 3*(3).

Parvin, R. H., & Salam, S. F. (2015). The effectiveness of using technology in English language classrooms in government primary schools in Bangladesh. *FIRE: Forum for International Research in Education, 2*(1), 47-59. http://preserve.lehigh.edu/fire/vol2/iss1/5

Potosi, L., Loaiza, E., & Garcia, A. (2009). *Using video materials as a teaching strategy for listening comprehension.* http://repositorio.utp.edu.co/dspace/bitstream/handle/11059/1936/371333A786.pdf

Ramin, V., Reza, B., & Nazli, A. (2014). The pedagogical utility of audio-visual aids on extrovert and introvert Iranian intermediate EFL learners' speaking ability. *International Journal of Research Studies in Educational Technology, 3*(2), 41-50. https://doi.org/10.5861/ijrset.2014.744

Ramírez, M. G. (2012). *Usage of multimedia visual aids in the English language classroom: a case study at Margarita Salas secondary school.* Unpublished master thesis. https://www.ucm.es/data/cont/docs/119-2015-03-17-11.MariaRamirezGarcia2013.pdf

Ranasinghe, A. I., & Leisher, D. (2009). The benefit of integrating technology into the classroom. *International Mathematical Forum, 4*(40), 1955-1961.

Şahin, M., & Şule Seçer, Y. E. (2016). Challenges of using audio-visual aids as warm-up activity in teaching aviation English. *Educational Research and Reviews, 11*(8), 860-866.

Tarnopolsky, O. (2012). *Constructivist blended learning approach to teaching English for specific purposes.* Versita. https://doi.org/10.2478/9788376560014

Viswanath, P. C., & Maheswara, R. C. (2016). The role of audio visual aids in teaching and learning English language. *International Journal of Scientific Research, 5*(4), 78-79.

Wilson, C. C. (2001). Visuals & language learning: is there a connection? *The weekly column, 48.* http://www.eltnewsletter.com/back/Feb2001/art482001.htm

Wright, A., & Haleem, S. (1991). Visuals for language classroom. Longman Group UK.

Yunus, M., Salehi, H., & John, D. S. A. (2013). Using visual aids as a motivational tool in enhancing students' interest in reading literary texts. In A. Zaharim & V. Vovovozov (Eds), *Recent advances in educational technologies* (pp. 114-117). WSEAS Press.

3 Dynamic and complex system approach to needs analysis, course development and evaluation of LANSOD courses in a French musicology undergraduate programme

Aude Labetoulle[1]

Abstract

When trying to analyse a LANSOD (LANguage for Specialists of Other Disciplines) training course, the elements that have to be taken into account are numerous and complex, and many questions are raised. For example, in the case of the English course in the undergraduate programme of musicology at the University of Lille SHS (France), how can high absenteeism be accounted for? What about the students' lack of motivation, or teachers' dissatisfaction at teaching the course? The purpose of the action-research project reported on is to understand and articulate these various elements so as to conceive, set up and evaluate a coherent language course, adapted to a specific context. We decided to adopt a dynamic and complex system approach as it was believed to be helpful in apprehending the complexity of a LANSOD context, guiding a needs analysis, and designing and evaluating a training course. We favoured the triangulation of sources, methods, and qualitative analyses in order to identify the main problems of the existing training course and to set the objectives of a new course. These results led us to conclude that a blended course with relatively highly specialised content was the most suitable option for this learning environment.

Keywords: systemics, needs analysis, course development, LANSOD, musicology, blended learning.

[1]. Université Lille SHS, Lille, France; aude.labetoulle@univ-lille3.fr

How to cite this chapter: Labetoulle, A. (2017). Dynamic and complex system approach to needs analysis, course development and evaluation of LANSOD courses in a French musicology undergraduate programme. In C. Sarré & S. Whyte (Eds), *New developments in ESP teaching and learning research* (pp. 31-50). Research-publishing.net. https://doi.org/10.14705/rpnet.2017.cssw2017.744

Chapter 3

1. Introduction

Most of the questions and topics which have dominated English for Specific Purposes (ESP) teaching and learning research – objectives of ESP programmes, task-based language teaching, computer-assisted language learning, needs analysis, materials development, etc. (Sarré & Whyte, 2016) – should all be taken into consideration when designing a LANSOD[2] course in a university context. Yet, despite scientific progress, the practitioner is still left struggling with an intimidating number of questions. For example, how can students' lack of motivation, or teachers' dissatisfaction with the course, be dealt with? How to cope with high absenteeism?

The purpose of the action-research project presented in this chapter is to understand and articulate these various elements so as to conceive, set up, and evaluate a coherent language course – in the specific context of the LANSOD English course of the undergraduate programme of musicology at the University of Lille SHS (France). We set up and evaluated the course over three cycles corresponding to three semesters. The analysis of the quantitative and qualitative data gathered when evaluating Cycle 1 will be presented in this chapter.

2. Framework of the action-research: the dynamic and complex system approach

2.1. Theoretical framework

Our attempt to conceive a useful tool for the analysis of a LANSOD course is based on research conducted in epistemology (Durand, 2013; Le Moigne, 2012; Morin, 1999, amongst others) and language learning and acquisition (Bertin, Gravé, & Narcy-Combes, 2010; Larsen-Freeman & Cameron, 2008; Waninge, Dörnyei, & De Bot, 2014). We selected seven key concepts to

---

[2]. The English acronym was coined by van der Yeught (2016) to translate the French acronym (LANSAD) originally coined by Michel Perrin. LANSOD refers to language classes destined to students whose major is not languages, but other disciplines such as medicine, chemistry, etc.

articulate what has been termed the "dynamic and complex system approach". The LANSOD course is considered a system in the sense that it is made up of elements which interact with one another to form a whole. These elements may be discrete, such as students, teachers, materials, input, etc., or more abstract, such as objectives or evaluation guidelines. The system is complex because its components (learners, input, etc.) and their interactions are complex. For example, what are the various interactions at play which account for learner motivation? A system is dynamic when it consists of processes that evolve over time, such as the language learning process. Because we are focusing on learning environments, the purpose of the LANSOD course is language learning, and it is assumed that there are several ways to reach this objective. The course is open in that it constantly interacts with a wider environment, for example the institutional context in which the class takes place. Lastly, this approach forces us to accept that a learning situation is so complex that ultimately we cannot understand it fully.

This approach is regularly adopted in language didactics (Mompean, 2013; Montandon, 2002; Waninge et al, 2014). One of its most accomplished interpretations is didactic ergonomics as developed in Bertin et al., 2010. Schematically, while fully acknowledging the dynamism, complexity and incertitude at play, the authors propose a five-pole model of the learning situation (learner, teacher, language, context, and technology), an in-depth analysis of the possible relationships between these poles, as well as possible pedagogical implications.

The model used in this study is an adapted and simplified version of the original model (Figure 1). The contexts include a professional context (depending on the students' future occupations) and an academic context divided into three levels – the macro (policies at the European and French levels), the meso (policies of the University of Lille SHS), and the micro (the musicology undergraduate programme). The learners' pole refers to all the students studying English in this programme, the teachers' pole refers to the teachers who have been teaching these students, and the contents' pole to what has been taught. The technology pole was integrated in the original context pole.

Figure 1. Simplified version of the didactic ergonomics model

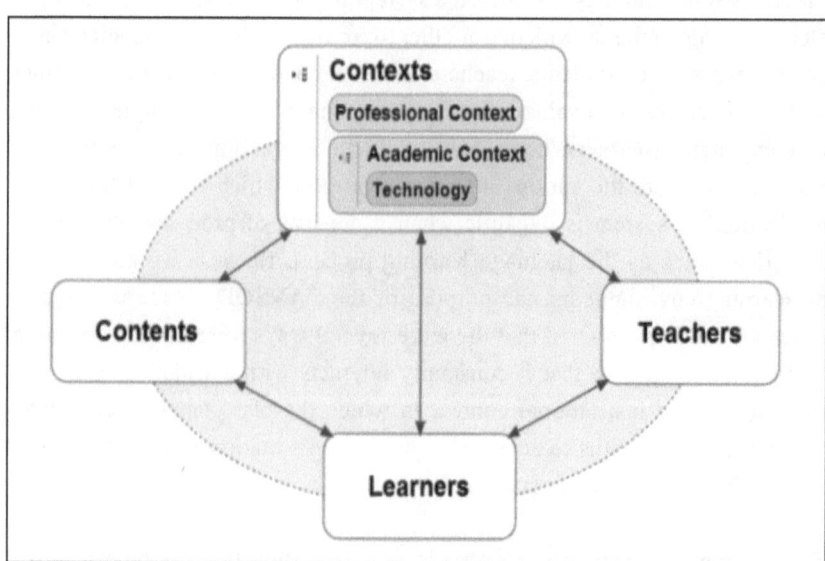

This conception of a LANSOD course has epistemological and practical implications. While it may seem relevant to have a general understanding of a language learning situation, it can quickly prove problematical when this raises more questions than it answers. Which elements should we take into account in the analysis of the course? How can we take its dynamic dimension into consideration? How can we design a new, appropriate course when we cannot understand everything of the general context? Ultimately however, it is precisely one of the strengths of this approach, because it helps reveal key, underlying, epistemological questions and thus encourages us to explicitly state our theoretical stance before undertaking the actual action-research. This process is all the more important as we are participants in the study. Therefore, we argue that it is crucial for us to deconstruct and report on every step[3] with "epistemic distancing" (Narcy-Combes, 2002). Efficiency is the main gauge of scientific validity; we try to devise appropriate actions which are the results of

---

3. In this respect, a more thoroughly detailed analysis of the study will be available at the end of the three-cycle research [2018/2019]

constructed (and deconstructible), informed choices. From a more didactic and pedagogical point of view, the structuring of the learning situation into poles and interactions also enables us to structure our future analyses, identify research questions, and interpret results; as will be seen later on, it also helps us organise the construction and the evaluation of the language course.

2.2. Methodological framework

The action-research project reported on consisted of five steps, as shown in Table 1.

Table 1. The five steps of the action-research project

Sept. 2015-Aug. 2016	Sept. 2016-Dec. 2017
1. pre-analysis of the existing learning environment to select our research questions for the needs analysis;	4. implementation of the course over three cycles;
2. needs analysis to identify the objectives of the new course and the means to reach our goals;	5. final evaluation (at the end of the study).
3. conception of the new course;	

As far as the needs analysis is concerned, our methodology was based on recent reviews of language needs analyses and procedures of data collection (Cowling, 2007; Long, 2005; Serafini, Lake, & Long, 2015). In keeping with our theoretical framework, we favoured an exploratory approach "so as not to preclude the possibility of discovering needs the needs analyst might not have considered" (Serafini et al., 2015, p. 13) with the use of open-ended questionnaires and interviews[4].

Triangulation of the data was done "to increase reliability and validity, [as] data should ideally be collected from two or more sources using two or more methods" (Serafini et al., 2015, p.12). We obtained data from five groups of participants (Table 2).

---

4. Learner questionnaire for the Needs Analysis ("Questionnaire – Music and Musicology Undergraduate Students"); translated from French to English; available at https://research-publishing.box.com/s/l3oba7fd0lff83naoui792tabgoh92gs

Chapter 3

Table 2. Participants in the needs analysis

Informants	Data collection methods
2 language supervisors	Questionnaires + interviews (recorded and transcribed)
2 content supervisors	
4 English teachers	
41 current students	Questionnaires
6 former students of the undergraduate programme	
Published literature (official publications from the Council of Europe, evaluation reports of the university, etc.)	

Data analysis then consisted in filtering our corpus thematically around each of the four poles, as illustrated in Figure 2.

Figure 2. Thematic analysis of the corpus

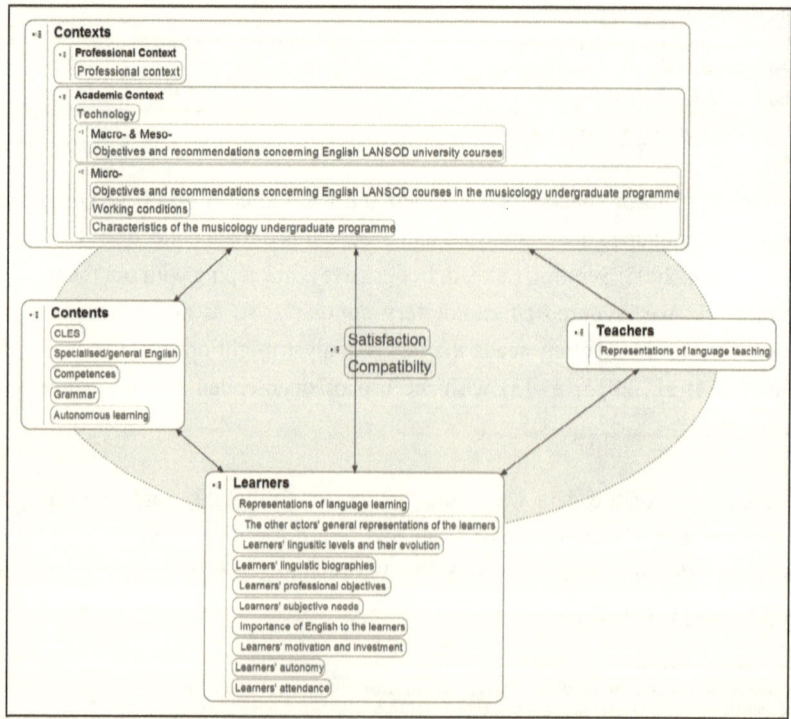

We used Excel (version 14.0.7177.5000, Microsoft) and Sonal (version 2.0.77, Alber) to filter and analyse our quantitative and qualitative data. The characterisation of the poles then enabled us to identify the main problems of the existing course the researcher could hope to address and the objectives of a re-designed course.

2.3. Evaluative framework

We implemented both continuous and final assessments. Formative assessment of the re-designed course was used to orient possible changes in the course, based on the feedback of the various participants during the cycle. Summative assessment, of which the data will be presented in this chapter, was aimed at evaluating the language course in relation to the objectives we had set. It took place at the end of each of the three research cycles via questionnaires completed by the learners and the English teachers[5]. The questionnaires yielded both quantitative data (with the use of Likert scales) and qualitative data (with the possibility of commenting on one's ratings).

3. Results from the needs analysis

3.1. General presentation of the undergraduate musicology programme

For all the musicology students attending the LANSOD English classes, it was compulsory to take a second language and the course was worth three credits, just like any other course. There were 12 two hour language classes in each of the six semesters of the undergraduate degree programme. Eight teachers taught two groups of first-year students (Y1) with 37 and 25 students, one group of 42 second-year (Y2) students, and one group of 46 third-year students (Y3). When asked whether they were satisfied with the classes, the learners rated their English classes 2.3 (out of 5) on average, with clear differences

---

5. Teacher questionnaire for the Evaluation of Cycle 1 ("Questionnaire – Cycle 1 Teachers"); translated from French to English; availbale at https://research-publishing.box.com/s/wm1li12dgirwxpiq7rvc0xd1uqu6wsps

between Y1 (3.3) and Y2 (1.5), while Y3 students gave the course an average rating of 2.5. These results echoed those of the former undergraduate musicology students among whom five out of six respondents declared being "rather unsatisfied" with their LANSOD classes. Teachers' satisfaction with their work varied greatly, with ratings ranging from 1 to 5 (out of 5). Two teachers considered not teaching these courses again because they disliked the experience. While more than half of the students believed they had reached the expected level at the end of the academic year (B1 in Y1 and B2 in Y3), even more students (22/38) declared that they felt they had stagnated or regressed since starting university.

3.2. Explanatory factors

We identified ten main factors, related to the four poles, which could account for low satisfaction rates and students' low sense of progress.

3.2.1. The 'context' pole

At a contextual level, we argue the objectives of the LANSOD courses were not sufficiently clear to all actors, mostly because of contextual restructuring. At the macro-level, "European language policy could […] be proactive and explicit on the basis of precise criteria […]. However, its application in schools is not automatic and requires a rethinking of the methods of language learning/teaching in universities"[6] (Mompean, 2013, p.32). The credit system changed our perception of personal investment as credits are "no longer calculated in relation to one hour of face-to-face class, but in relation to the student's personal workload which is thus given more importance than before" (Mompean, 2013, p. 25). In the meso-context, restructuring was underway at several levels. The impact on LANSOD classes of the upcoming merger of Lille SHS with two other universities (2018) was still to be determined. In 2012, the creation of a 'LANSOD department' was aimed at defining common objectives and promoting a better overall coherence; in 2015-2016, numerous

---

6. Author's translation

debates were still going on about grouping students by ability or proficiency level (known as 'tracking'), innovation and administrative restrictions, 'liberté pédagogique' (pedagogical freedom), and certifications. At the micro-level, the decision-making structure was not completely clear. The structuring of LANSOD courses within the arts department, of which the undergraduate musicology programme is part, was still 'ongoing'; someone was appointed but their role (supervisor? coordinator?) was not clearly defined as of June 2016. Because all actors put forward the principle of 'liberté pédagogique', it was hard to tell what was solely a 'recommendation' or a 'rule' when it came to the content supervisors' and language coordinators' instructions. This ambiguity was well encapsulated in the terse description of the language courses in the musicology undergraduate programme guide of studies: 'UE 9, Languages'. As far as working conditions go, they were rated 2.5 (out of 5) on average by teachers, the most negative elements being the number of students per group, absenteeism, lack of technological equipment in classrooms, mixed language proficiency levels, and sometimes challenging collaborative work with the administration.

3.2.2. The 'teacher' pole

The teaching team was also being put together. Seven out of the eight English teachers had not previously taught this class and only two eventually continued teaching the following year, which makes a high turnover rate. They had various backgrounds and experiences. Out of the two teachers with tenure, the two contract teachers, and the four teaching assistants, five had little or no prior experience of teaching at university level. Their majors included ESP, literature, translation, and French, in France or abroad. They were generally interested in music, but very few (1/8) were familiar with the domain of musicology, let alone ESP in this area. Collaborative work was encouraged by the coordinator but the attempts were seen as timid and not always successful, because of teaching methods deemed too different, and the lack of time and will to communicate. As far as information and communications technology is concerned, teachers were quite positive about its use in the classroom, but some underlined that they did not really take advantage of all its possibilities.

Chapter 3

3.2.3. The 'learner' pole

Students' language levels ranged from A1/A2 to C2 and the degree of importance they attributed to English varied depending on the context. If 59.5% of students considered it important or very important in general, 59.4% considered English for their studies and their professional lives unimportant or not very important. The tasks they mentioned concerning English in their studies and in their personal lives were overall the same (reading specialised articles and talking with artists, understanding lyrics of a song, traveling, watching movies and series, playing videogames, and using the Internet). However, the examples varied considerably as regards English for their professional lives. These results came to little surprise, as their professional objectives were quite diverse, and therefore so were their needs as regards English. Music teaching stood out, but as it is carried out in varied contexts, we could infer the needs of the students would not be the same (e.g. music teachers in conservatories as opposed to primary school teachers). It all confirmed data obtained from the former undergraduate musicology students. All respondents underlined English was important, but to various degrees and depending on the domain; the needs of a music teacher in a French middle school were quite different from those of an instrumentalist working abroad.

Half of the teachers declared that absenteeism and the lack of investment were some of the main causes for their dissatisfaction at teaching the course, as it made it difficult to know what to expect and virtually impossible to create a coherent programme with a progression. Attendance at university is not compulsory, except for evaluations. On average, students declared that they had been in class more often than not (55%), but most teachers indicated hardly ever having more than half of the students in Y2 and Y3, sometimes even just one or two students. The students rarely in class (19/41) stated that the two main reasons for not attending the courses were disinterest in the course (11) and timetable constraints (3). The content supervisors added that some students needed to work on their instruments several times a day, while language teachers also blamed the working conditions (large groups and no projectors). If for a class which counts for three credits students are expected to do 60-75 hours of work outside class time, students declared on average having spent 80 hours on English in total, but their answers varied greatly (from 20 to 450 hours) and only 23.3% of them answered the question, when the other 76.6 % stated they had no idea, did not answer the question, or indicated they had only worked just prior to the evaluations. In total, 59% of students declared the amount of work they had provided was unsatisfactory. Sixty-one percent said they never (25%) or rarely/ sometimes (36.1%) used English out of the classroom.

Paradoxically, all students acknowledged that personal work is very important in language learning, and when asked how much time per week they should spend on it, they responded about 2 hours on average. We argue that two key elements accounting for their lack of investment were their level of autonomy, half of them declaring not to be autonomous learners (11/22), and lack of motivation. Their motivation levels in the English classroom either dropped (16/36), depended on the semester (8/36), were stable (7/36), or even increased (5/36) over time.

The most important motivating factors were the marks, according to the content supervisor, as well as the use of content linked to music. The learners specified what would motivate them more, but their answers were so varied that no real consensus emerged. Their suggestions included more content linked to music, same proficiency groupings, more grammar, less homework, smaller groups, and better timetables. The quality of student-teacher relations was also important to students; when asked whether the teachers had met their expectations, 40% of respondents said they had not, 40% were satisfied and 20% said it depended on the semester.

3.2.4. The 'content' pole

The various actors of the learning environment had sometimes different views of what the objectives of the LANSOD courses should be. Our analysis showed there was a weak consensus about the importance of language learning, the professional dimension of university classes, the notions of threshold levels, communicative competences, and tasks. Working on all language skills,

Chapter 3

autonomous learning, and the need to adapt to the diversity of the public were not objectives shared by all. Most teachers were not opposed to coordinating methods and content as long as pedagogical freedom remained paramount. There was no consensus concerning tracking (i.e. grouping students by proficiency level) or about the content of the LANSOD undergraduate musicology classes: how much importance should be given to specialised content linked to music? To the CLES[7]? And to the notion of "getting by abroad" frequently mentioned by learners?

3.3. Conclusion of the needs analysis

The needs analysis was complex to carry out and we obtained limited data. The response rates were rather low, especially among the learners (learners: 41/150; language teachers: 4/8; language supervisors: 2/3; content supervisors: 3/3; former musicology students: 6/110). Because we had to resort to limited convenience samples rather than stratified random samples (Long, 2005, pp. 34-35), the representability of the samples is questionable. Our methodology could also have been more rigorous (as illustrated in Serafini et al., 2015 for example): there was no sufficient pilot testing of the questionnaires, they were submitted to the participants late and there were too many open-ended questions, which made the analysis more complicated.

Overall, however, the difficulties we encountered seem to be quite common in needs analysis (Serafini et al., 2015, p. 24) and these limitations were put into perspective if we bear in mind that our purpose was to obtain a general understanding of the learning situation adequate enough to enable relevant decision making. Indeed, the needs analysis helped us identify some inconsistencies between the various contextual levels as well as a lack of horizontal coherence at a micro-level, all this impacting satisfaction levels and students' perceptions of progress in language proficiency. When designing the new courses, these are problems we will have to try to solve.

---

7. Certificat de Compétences en Langues de l'Enseignement Supérieur, a language certificate at university level based on the Common European Framework of Reference (CEFR) for languages, created in 2000

4. Design of the new course

4.1. General objectives and evaluation

We were then able to infer potential objectives, means to reach these objectives, indicators, and assessment tools of the new course[8]. We decided to focus on Y1 and Y3 for Cycle 1 of the research, as we lacked the time to coordinate with several teachers before term started. Even though we could not change the contexts, the teachers, nor the learners directly, we could have a direct impact on the content of the courses as well as the general teaching methods. From the results of the needs analysis, we inferred that a blended course with relatively highly specialised content would be the most suitable option.

4.2. Contents

The syllabus mainly focused on disciplinary components, with a gradual transition towards more specialised English from Y1 to Y3. Unfortunately, due to time constraints, we could not carry out a proper discourse analysis of English for Music, but our choices were however informed by data collected from content specialists, former students of the programme, as well as the learners. Based on how frequently some communication situations were mentioned and the feasibility of transforming them into language learning objectives, we organised the syllabus around tasks.

In Y1, the semester was organised around the topic of music festivals. In groups, learners had to present a project of a music festival to sponsors and write its programme. In Y3, the main theme was one's instrument; learners had to write an ad to sell their instrument, improvise when asked about their instrument and their practice, as well as read part of a score in English. Enabling the students to take the CLES exam also became one of the explicit objectives of the course, and all five language skills were therefore focussed on. Deciding to aim for the CLES, as well as taking into account the diversity of the learners' needs, led us to

---

8. For an overview of the objectives and the indicators, see the first two columns in Table 2.

Chapter 3

add general topics to the syllabus: each class started with students summarising the news of the week, and we developed activities focussing on the American presidential elections. In order to address the individual needs of learners and to encourage autonomous learning, we introduced Personal Projects. They consisted in learners individually choosing which competence they wanted to work on, devising a plan to work on that competence, submitting their work regularly to the teacher for feedback during the semester, and presenting the work to the class at the end of the semester.

4.3. Methods

We decided to work on this content in a blended environment. The face-to-face classes and the online modules were organised around tasks. To try to make the best use of each mode, face-to-face classes were focussed on production (both oral and written), whereas online modules were devoted to receptive skills. In all, there were five online modules and seven face-to-face classes.

Figure 3. Organisation of the blended course

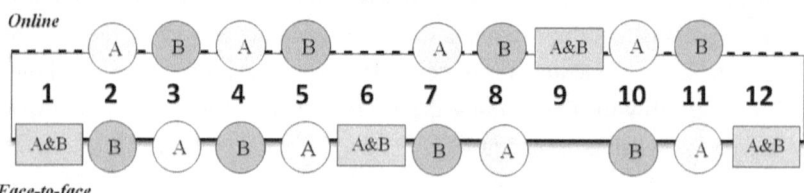

The learners of each class were therefore divided in two smaller groups (groups A and B) based on language proficiency as assessed after a placement test at the beginning of the academic year. The online modules (top line in Figure 3) were made available on a Moodle platform and all online activities were marked to encourage learners to be exposed to English at least once every two weeks if they did not come to face-to-face classes (bottom line in Figure 3), and to enable teachers to create a real progression in their course, regardless of how many and which students came to class. The courses (comprising 12 sessions in total – both face-to-face and online – as represented in Figure 3) in this respect were quite restrictive, but the students had 14 days to cover the online content. Online modules typically consisted of auto-corrective activities including a vocabulary test, an oral and/or written comprehension activity with preparation exercises, and a grammar lesson with exercises. Teacher support was considered paramount, so each module ended on a feedback activity which was discussed in face-to-face classes. The face-to-face classes were also described on Moodle and students who had missed a class could download all the materials used in class. The evaluation was in keeping with the instructions of the language coordinator, the learners' preferences and our objectives:

- Y1: In-class written exam (35%) + oral presentation and written group exercise (25%) + online sessions (25%) + personal project (15%)

- Y3: Two in-class exams (55%) + online sessions (25%) + personal project (20%).

5. Results of the evaluation of the re-designed LANSOD course (Cycle 1)

5.1. Results

Table 3. Findings from Cycle 1

Poles	Objectives of the new course	Assessment indicators	Results
All poles	Obtain a higher satisfaction level from learners and teachers	Learners' and teachers' satisfaction and motivation levels	Learners = ✓ Teachers = ✓
	Foster linguistic development and obtain better results from students' self-assessment	• Students' self-assessment of linguistic progression • Teachers' assessment of linguistic progression	Self-evaluation from learners = ✓ Teachers = ! (room for improvement)

Chapter 3

All poles	Propose a clearer definition of the objectives to all actors based on the needs analysis	Adequacy of the perceived objectives by learners, teachers and direct supervisors	Adequacy learners/ teachers = ! Content supervisors = ✓
Teachers	Reduce teachers' workload and propose a coherent and flexible three-year syllabus	• Teachers' workload • Degree of collaboration • Reduced workload and re-use of teaching materials	X & ✓
Context	Improve working conditions for learners and teachers	Learners' and teachers' satisfaction levels concerning the number of students per group and same proficiency grouping	✓
Learners	• Propose learning and teaching conditions which foster qualitative relationships between learners and teachers • Take into consideration the diversity of learners' needs in the design of the syllabus	Adequacy of learners' needs with the syllabus	Learners = ✓ Teachers = ! & ✓
	Encourage personal work outside the classroom	• Time spent by students doing English outside of the classroom • Regularity of learners' work • Learners' satisfaction as regards personal investment	✓
	Guide the learners towards more autonomy	• More autonomous learners • Changes in learners' use of tools (e.g. using dictionaries more sensibly) and learning strategies	✓
	Motivate learners	Levels of motivation	✓
	Obtain an attendance rate which is satisfactory to teachers	Attendance rate	✓

Two teachers (Teacher A with one group in Y1 and the group in Y3, and Teacher B with the other group in Y1) and 98 students were involved in Cycle 1. An overview of our findings from Cycle 1, in which the methodology described in 2.3 was used, is provided in Table 3 above.

The results concerning the two main objectives of the new course were rather positive. Learners gave an average rating of 3.7 out of 5 on a Likert scale to describe their satisfaction, a clear improvement compared to the previous rating (2.3). Teachers' ratings were more homogenous than before and also improved, with an average of 3.3. This time, 73% of learners considered themselves as having progressed or progressed a lot. Teacher A was in charge of designing the class and materials, and considered their workload excessive as they estimated they had worked an average of 15 hours a week to prepare four hours of class. However, they took a long term view, knowing that the materials would be re-used. Teacher B only provided occasional support and considered their workload satisfactory. The teachers were frequently in contact: 66 emails were exchanged and four short informal meetings were held. The learners were overwhelmingly positive about same proficiency grouping, rating this aspect 4.4 out of 5. Teacher B gave it a rating of 5, and Teacher A a 5 in Y3 but only 3.5 in Y1, considering it was sometimes difficult to teach a group with no class leaders. When asked whether they believed the courses were adapted to their needs, 71% of learners rated it 3.5 or above, 78% rated the content of the syllabus 4 and above, and 80% 4 and above as regards the competences included in the syllabus. The learners with lower scores had generally had problems accessing the online content[9]. The teachers rated the course 3.6 when assessing the appropriateness of the course to their students' needs; they shared the concern that there had been too much content and that, for the lower proficiency Y1 group, the objectives were slightly too demanding. For Y3, Teacher A also feared some learners might have found the workload and the system too demanding and restrictive. Concerning learners' investment and motivation, the results were also positive. On average, learners said they had worked 32 hours, including both face-to-face and online sessions. We are short of what could have been expected, but 56% of them were satisfied

---

9. About one fourth of students in Y1 only enrolled officially in the university one month after the beginning of the school year due to administrative problems and could therefore not have access to the Moodle platform over that period of time.

with their involvement. There was also marked improvement in the regularity of the learners' work: 60% declared that they had worked regularly, as opposed to only 22% previously, and 83% of the learners rated their motivation 3 and above. As for autonomy, 67% considered they were more autonomous than before. The teachers agreed that learners seemed more autonomous thanks to the personal project and the blended learning system. Concerning absenteeism, 69% declared they had come to all or all but one class. To finish, the learners indicated that the teachers had met their expectations (97%).

5.2. Consequences for Cycle 2

Based on these results, the course in Cycle 2 will offer the same balance between general and specialised English. More attention will be paid to grammar in face-to-face classes with the lower-proficiency Y1 group, as the learners indicated dissatisfaction in this area. We will attempt to encourage more production in face-to-face classes as the activities of Cycle 1 were too ambitious and left little time for student production. More choice will be given in Y3 with the selection of some content of the syllabus by students. Particular attention will be devoted to the online modules, considered as the weak point of the new course as they were rated 2.7 on average by learners. It can be accounted for by the difficulty for Y1 students to access the modules and some technical problems and human errors (e.g. spelling mistakes in auto-corrective exercises). We will shorten the classes and make the structure of the course clearer to learners as in the feedback sections the majority of learners found the online modules too long and some learners were sometimes confused as to when to come to class. To finish, more collaborative work will be introduced by Y2 teachers, at least when it comes to the definition of the objectives of the course.

6. Conclusion

The results indicate the new course helped us reach the majority of our objectives. Even though this action-research project is not over, we can already state that the framework and the methodology we adopted to re-design and evaluate the

LANSOD course were highly instrumental in its success. The needs analysis, which was structured around the theoretical and methodological guidelines of the dynamic and complex system approach, helped us characterise the existing learning environment, identify the main problems, and then make appropriate decisions. It led us to conceive a course whose key aspects were specialised content, the articulation between face-to-face classes and online modules, and the personal project. At the end of the action-research cycle, we hope to assess the transferability of our theoretical and methodological framework, as well as establish which pedagogical components could be easily transferable to other LANSOD courses (such as the blended system and the personal project).

References

Bertin, J.-C., Gravé, P., & Narcy-Combes, J.-P. (2010). *Second language distance learning and teaching: theoretical perspectives and didactic ergonomics.* IGI Global. https://doi.org/10.4018/978-1-61520-707-7

Cowling, J. D. (2007). Needs analysis: planning a syllabus for a series of intensive workplace courses at a leading Japanese company. *English for Specific Purposes, 26*(4), 426-442. https://doi.org/10.1016/j.esp.2006.10.003

Durand, D. (2013). *La systémique: «Que sais-je?» n° 1795.* Presses Universitaires de France.

Larsen-Freeman, D., & Cameron, L. (2008). *Complex systems and applied linguistics.* Oxford University Press.

Le Moigne, J.-L. (2012). *Les épistémologies constructivistes: «Que sais-je?» n° 2969.* Presses Universitaires de France.

Long, M. H. (2005). *Second language needs analysis.* Cambridge University Press. https://doi.org/10.1017/CBO9780511667299

Mompean, A. R. (2013). *Le centre de ressources en langues: vers la modélisation du dispositif d'apprentissage.* Presses Universitaires Septentrion. https://doi.org/10.4000/books.septentrion.16720

Montandon, C. (2002). *Approches systémiques des dispositifs pédagogiques: enjeux et méthodes.* Editions L'Harmattan.

Morin, E. (1999). Organization and complexity. *Annals of the New York Academy of Sciences, 879*(1), 115-121. https://doi.org/10.1111/j.1749-6632.1999.tb10410.x

Narcy-Combes, J.-P. (2002). Comment percevoir la modélisation en didactique des langues. *ASp, (35–36)*, 219-230. https://doi.org/10.4000/asp.1678

Sarré, C., & Whyte, S. (2016). Research in ESP teaching and learning in French higher education: developing the construct of ESP didactics. *ASp, (69)*, 139-164. https://doi.org/10.4000/asp.4834

Serafini, E. J., Lake, J. B., & Long, M. H. (2015). Needs analysis for specialized learner populations: essential methodological improvements. *English for Specific Purposes, (40)*, 11-26. https://doi.org/10.1016/j.esp.2015.05.002

Van der Yeught, M. (2016). *Developing English for specific purposes (ESP) in Europe: mainstream approaches and complementary advances.* Sub-plenary lecture, ESSE conference, Galway, Ireland, 23 August 2016.

Waninge, F., Dörnyei, Z., & De Bot, K. (2014). Motivational dynamics in language learning: change, stability, and context. *The Modern Language Journal, 98*(3), 704-723. https://doi.org/10.1111/modl.12118

4 Designing and implementing ESP courses in French higher education: a case study

Susan Birch-Bécaas[1] and Laüra Hoskins[2]

Abstract

This chapter reports on the design, implementation and evaluation of an English for Specific Purposes (ESP) course for dental students at the University of Bordeaux. We give an overview of the 'English for Dental Studies' courses taught from second year through to fifth year before focussing on the fifth year course in which the students' task is to present a case treated on clinical attachment. By following the schema of Cheng's (2011) 'basic considerations', we will briefly describe the process from needs analysis and identification of learning objectives, to designing materials, learning tasks, and assessment criteria, with a focus on methodologies. Feedback from students via questionnaires is analysed in order to compare their perceived needs and expectations pre-course with their impressions after the course. Finally, we explore the gains that can be made by both ESP specialists and disciplinary teachers in the context of internationalisation in the French higher education system.

Keywords: ESP, course design, CLIL, feedback, tasks.

---

1. Université de Bordeaux, Bordeaux, France; susan.becaas@u-bordeaux.fr

2. Université de Bordeaux, Bordeaux, France; laura.hoskins@u-bordeaux.fr

How to cite this chapter: Birch-Bécaas, S., & Hoskins, L. (2017). Designing and implementing ESP courses in French higher education: a case study. In C. Sarré & S. Whyte (Eds), *New developments in ESP teaching and learning research* (pp. 51-69). Research-publishing.net. https://doi.org/10.14705/rpnet.2017.cssw2017.745

Chapter 4

1. Introduction

ESP has traditionally been considered as a "practitioner's movement" (Johns, 2013, p. 6) focussing on learner needs and pedagogical applications and Hyland (2013) has referred to it as "research-based language education" (p. 107). Learner needs are established by discourse analysis, genre analysis, and study of professional communities, with much work carried out on identifying the rhetorical and linguistic characteristics of various types of specialised discourse and on describing the way in which different discourse communities function (Hyland, 2013; Swales, 1990). In France, ESP has often been separated into two strands: a teaching strand (*LANgues pour Spécialistes d'Autres Disciplines*) and a discourse strand (*anglais de spécialité*), with the latter drawing more attention from research communities. However, Swales (2011) has argued that "we have had, over the 50 year history of ESP, all too little careful research in what actually happens in our classes" (p. 273). Belcher (2013) confirms that "some in ESP might well argue that the community that ESP professionals know the least about is its own" (p. 544). Descriptions of course design and material development are often dated and as new courses are put in place, it could be argued that ESP specialists do not always take the time to reflect on their practice and build on this experience within the frame of action research. Belcher (2013) points to this lack of analysis when she questions "How do ESP specialists know that what they do results in the learning outcomes that they and their students desire?" (p. 544). It may be that student needs, and course design and materials, have been less researched than specialist discourse yet both domains are closely linked and indeed the former stems from the latter. Johns (2013) refers to the early work of Dudley-Evans and St John (1998) where the key roles of the ESP specialist are described as being teacher, course designer, materials provider, collaborator (with subject specialist), researcher, and evaluator. Here, we view an ESP course from these multiple perspectives. Indeed, ESP course design goes hand in hand with research as the ESP specialist assesses needs, analyses target genres and language use in the community of practice and designs appropriate materials from specialised corpora which will draw attention to certain linguistic conventions. While it has been pointed out that not all ESP teachers are prepared or trained for this (Belcher, 2006; Van der Yeught, 2010; Wozniak, Braud, Sarré,

& Millot, 2015), in the *Département Langues et Cultures* (DLC) there is a strong ethic of team-teaching, reflective practice, and research-driven pedagogy, enabling ESP novices to 'train on the job' and encouraging more experienced ESP teachers and ESP researchers to share their expertise.

ESP courses traditionally begin at undergraduate level with general English for academic purposes or study skills and then move to more subject-specific conventions as students acquire more disciplinary expertise. Hyland (2013) describes the ESP teacher's role as "identifying the specific language features, discourse practices and communicative skills of target groups" (p. 6). This begs the question then of how much actual domain-specific expertise is required of ESP teachers? Indeed, this question was asked in the early days of ESP. Robinson (1991) argued that ESP teachers should not try to be 'pseudo-teachers' of subject matter and in the first volume of the French ESP journal, *ASp*, Tony Dudley-Evans (1993) entitled his article *Subject Specificity in ESP: How much does the teacher need to know of the subject?*. He comes to the conclusion that knowledge of a community, its discourse, and genres is more important than very specific content knowledge although the teacher obviously needs to take an interest in and be curious about the subject matter. As we will see below, the role of the ESP teacher in the 'English for Dental Studies' courses at the University of Bordeaux moves from providing disciplinary-related materials accompanied by scaffolding activities to increased collaboration with the subject specialists and investigation of more specific disciplinary discourse.

In this chapter, we aim firstly to explore how needs analysis can inform task design and evaluation formats to respond to what Hyland (2002) has termed the students' 'demand for personal relevance'. Wozniak and Millot (2016) have also emphasised the need for professional relevance and acquisition of a disciplinary and professional culture in English. We also focus on materials, activities, and tasks to explore how an ESP course can raise awareness of specialised language through noticing tasks (Ortega, 2015) and enable students to express their "already established disciplinary expertise" (Whyte, 2016, p. 14). Hyland (2011) points out that learners acquire features of the language as they need them and therefore this type of specific approach is more motivating. Finally, we also investigate whether a student's expertise in the discipline can influence their language competence. Whyte (2013) describes how the level and currency of content knowledge and its centrality in the life of the user can influence the development of the discourse domain. A questionnaire was thus used to gather the students' perceptions on how they had achieved the task and their opinion on other aspects of the course. Their answers also enable us to analyse how teacher feedback on performance can be provided without impairing the students' motivation and self-confidence.

2. Needs analysis

2.1. The context

The DLC at the University of Bordeaux provides English courses for a cohort of some 100 students admitted into the School of Dentistry after a first medical foundation year[3]. Dental studies are divided into three stages in France: the first undergraduate stage covers first to third year, the second postgraduate stage fourth to fifth year, and the third clinical stage sixth year and beyond (a maximum of four years). Ministry guidelines[4] stipulate that by the end of the undergraduate stage, dental students should be able to read and present scientific texts written in English and that they should attain a B2 level of competence according to the Common European Framework of Reference for languages (CEFR). At the beginning of their English course at the DLC, second year dental students take a language placement test[5]. Out of the 96 students who took the test in 2017, only 19% attained this level or above, indicating that the English programme should cater for their needs in general English as well as English for specific purposes. It should be noted that the students' levels in English were above the national average in France, where,

---

3. Première année commune aux études de santé (PACES)

4. http://www.enseignementsup-recherche.gouv.fr/pid20536/bulletin-officiel.html?cid_bo=71552&cbo=1

5. The Oxford Quick Placement Test

according to a European Commission's (2012) survey, only 14% of students attain a B1 level or above by the time they leave high school.

Before discussing the fifth year English course and the focus of this chapter in detail, it is necessary to situate it within the wider programme of courses in 'English for Dental Studies' that are provided in the second to fourth years of study. There is no provision for English in the first medical foundation year of dental studies during which students in France from across the health sciences take a cross-disciplinary competitive entry exam into medicine, midwifery, pharmacy as well as dentistry. Table 1 sums up the changing focus of the 'English for Dental Studies' courses at the University of Bordeaux. As students gain disciplinary knowledge and skills and move toward their future profession, the English courses thus progress towards more disciplinary and professional objectives. Whereas the second and third year courses aim at developing a broad range of communication skills, with the emphasis shifting from understanding and interacting in second year to expressing oneself at length in third year, the fourth and fifth year courses are project-based courses. In fourth year, students have the task of assembling a small corpus of research articles that respond to a specific problem encountered in clinical practice. They must read and review the literature before presenting it to an examining panel and their peers. This task runs parallel to a disciplinary course the students follow in their fourth year entitled *Lecture Critique d'Articles*, where they learn to read research articles critically. Finally, in fifth year, students present a clinical case that they have treated during hospital attachments. As we shall see later, these tasks were devised by the English teachers in collaboration with disciplinary lecturers.

The organisation of the 'English for Dental Studies' courses reflects this shift towards disciplinary competence and autonomy, with weekly structured contact hours in second and third year, divided between the classroom and the language centre, where resources and activities are tailored to individual learner profiles (there is no ability grouping for the classroom hours). In fourth and fifth year, students have fewer structured contact hours to allow them to work on their English projects in the language centre, and the contact hours they have are both in a classroom and tutorial setting, spread over the semester.

Table 1. ESP Courses from first to fifth year dentistry

	Organisation and Objectives	Tasks and Assessment	Class materials
Year 1	Medical foundation year; no English instruction		
Year 2	• 30-hour blended learning course • Developing communication skills for dentists • Interacting orally with disciplinary peers • Learning to learn	• Receptive skills test (reading, listening, grammar, vocabulary, pronunciation) • Continuous assessment of productive skills (written learning diary, oral interaction)	• Video and text (popular sources) provided by teacher with accompanying tasks and communicative scenarios
Year 3	• 30-hour blended learning course • Informing patients about a dental condition • Presenting a dental topic to peers • Discovering disciplinary resources	• Productive skills tests (written blog post and oral presentation) • Continuous assessment of productive skills (oral interaction)	• Video and text (popular sources) and disciplinary texts provided by teacher with accompanying tasks and communicative scenarios
Year 4	• 20-hour blended learning course • Using disciplinary texts to explore a problem encountered in clinical practice • Communicating on and discussing findings with disciplinary peers	• Productive skills test (oral presentation)	• Disciplinary texts/ figures provided by students and teacher. Accompanying tasks and communicative scenarios
Year 5	• 20-hour blended learning course • Reflecting on clinical practice • Telling the story of a clinical case • Discussing treatments with peers	• Productive skills test (oral presentation)	• Disciplinary texts (case studies and photos) provided by students and teacher • Accompanying tasks and communicative scenarios

2.2. Learning objectives (fifth year)

As we have just seen, the objective for fifth year students is to be able to present a clinical case to their peers and dentistry lecturers. To discuss the needs of the students for this new course, the team of ESP teachers met with the dentistry lecturers. This gave us the opportunity to take stock of the courses and materials used with second, third and fourth year students. We were also able to underline the specificity of the ESP courses and the complementarity of blended learning where time spent in the language centre could be dedicated to more personalised objectives, discussion workshops, cultural events, tandem pairings, activation or consolidation of specific language skills, general English, and Test of English for International Communication (TOEIC) preparation, etc.

Our colleagues from the school of dentistry were able to give us an insight into the hospital context and the students' work there. These expert members of the community gave their perception of the student's academic and professional needs and were able to draw parallels with the academic tasks which were required of the students in French that year (notably the 'CSCT'[6] oral exam in which students are given a case which they must analyse and present to their teachers). The consensus was that the students should work on tasks related as much as possible to their clinical practice as this was to be the main focus of their fifth year of study and seemed to be a logical progression from the tasks carried out in fourth year. Indeed, in the fourth year, when students were asked to mingle with their peers and recount 'an interesting/difficult/original/challenging case, etc.' seen at the hospital, we had noted their motivation and enthusiasm and this seemed to be what students enjoyed talking about most. The ability to discuss cases in an English lingua franca context or being able to present a case study at a conference are part of the students' disciplinary and professional socialisation. The dentistry colleagues agreed that the rhetorical and communicative skills gained in the ESP course might assist students in their French CSCT exam. At the same time, for the ESP specialists, it was hoped that this task-based approach would enable the students to draw on and express their disciplinary knowledge.

---

6. Certificat de Synthèse Clinique et Thérapeutique.

It was decided that the examining panel for the case study presentation would consist of both the ESP teacher and the dentistry lecturer and that a session of team-teaching with the dentistry colleagues presenting a case to the students would heighten the latter's motivation. In terms of timetabling English instruction, students would have to be allowed enough time in their already dense schedules to work on their English projects and hospital placements remained the priority. Consequently, we decided to see the students for two input sessions at the beginning of the course and then two tutorial sessions where the students, in pairs, could report on the progress made on their project and receive individualised feedback by rehearsing their presentation. A final input session was programmed in the weeks leading up to the final presentation. For the rest of their 20-hour course, students could work semi-autonomously in the language centre.

3. Materials, activities, and tasks

3.1. A genre approach

The objective of the first session is to familiarise the students with the top-down structure of a case study – the elements which are to be presented and the typical order in which they are found. The students thus examine the practices of their community and discover the conventions of the case study genre. The input is accompanied by activities, for example students work collaboratively to re-order several 'jumbled' case reports taken from the British Dental Journal and match them to their figures. Analysis of the different steps, rhetorical functions, or 'moves' can then be checked against a template provided by the dentistry teachers which advises students on conventions (Figure 1). Hyland (2015) has warned against 'constraining templates', but at this level the scaffolding provided enables students to structure their information, follow the norms of their community, and imitate these highly conventionalised productions. From this perspective, genre analysis "provides non-native speakers with the linguistic and rhetorical tools they need to cope with the tasks required of them" (Dudley-Evans, 1997, p. 62).

Figure 1. The typical macro-structure of a case study

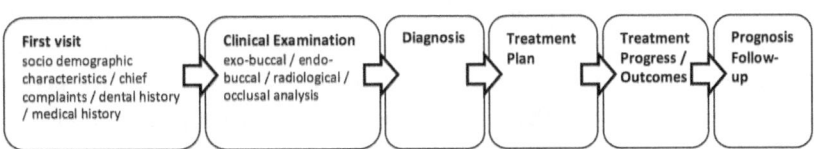

Now aware of the framework, the students need to analyse how these steps are achieved linguistically. They mine the mini-corpus of case studies for frequently occurring language clusters, searching for lexico-grammatical patterns and linguistic conventions used for the rhetorical functions (describing the patient's profile, giving the reason for referral, charting the medical history, etc.). Students work collaboratively to enter their findings into a shared file online, leaving by the end of the session with a lexicon for presenting a clinical case. The students examine and report back and this explicit analysis of examples contributes to raising awareness of certain patterns.

3.2. The written to oral register

After working on examples of written case studies, students are encouraged to think about how they could convey similar information orally to both dentists and non-dentists, as they will have to do in their final presentations. Students are given the task of sorting cards containing similar information into three categories of register: (1) what you might write in a scientific journal, (2) what you might say to another dentist, and (3) what you might say to a non-dentist (for one example, see Table 2). In their current and future practice, French dental students are most likely to find themselves in situations of the third type, in explaining treatments to non-Francophone patients. Due to the difficulty in accessing authentic examples of utterances of the last two types, we formulated oral alternatives to excerpts from the written case studies read in a previous class by the students. Based on observations made by Carter-Thomas and Rowley-Jolivet (2003) on the syntactic differences between the written genre of a proceedings article and the oral genre of a conference presentation, our suggested paraphrases include a higher frequency of active structures, personal pronouns, *there* structures, pseudo-clefts, and shorter, less dense syntactic chunks than are present in the written case reports. Our paraphrases for non-dentists endeavour to offer alternatives to medical jargon that lay people would understand. Once the students have classified the items according to register, they are invited to notice the syntactic differences between the three styles, which leads to a discussion. For some written excerpts, we offer no oral paraphrases, leaving cards blank for students to put forward their own suggestions once they have finished classifying the items we provide. In this way, students are sensitised to the differing styles of the oral and written register in English, and are given linguistic strategies to avoid 'talking like texts' during their final presentation.

Table 2. Register activity

what you might write in a scientific journal	what you might say to another dentist	what you might say to a non-dentist
The patient was prescribed a course of prophylactic antibiotics (amoxicillin 500 mg TDS and metronidazole 200 mg TDS) for one week and sent home.	We prescribed the patient a week's course of prophylactic antibiotics – amoxicillin 500 mg TDS and metronidazole 200 mg TDS, to be more specific. Then we sent her home.	We put the patient on antibiotics for a week to prevent any infection and sent her home.

3.3. Peer-to-peer instruction and feedback

In addition to raising awareness of the distinct language forms of the oral genre, the classroom sessions also encourage the students to reflect on the linguistic features that accompany another semiotic mode of oral presentations: visuals. For the communicative task at hand, students will have to walk the audience through their interpretation of X-rays and photographic images to help them see the salient parts related to treatment choices. To this end, we developed a series of communicative activities that elicit the language seen previously (Ortega, 2015) and give students a chance to practice and consolidate appropriate language forms while the teacher guides and facilitates interaction by monitoring the students as they complete the tasks with their peers. The group work also allows for peer feedback on subject content as students will comment and ask questions about each others' cases and indeed help each other with language difficulties. These activities also exploit resources provided by class members, who are as Belcher (2006) says "the most significant subject-area resources in an ESP class" (p. 172).

3.4. Phonology issues

Phonology issues are another essential aspect of oral presentations in the target language that the English course for fifth year dental students seeks to address. In the time available for the course, it would be over-ambitious to aim to remediate all the phonological difficulties encountered by our French-speaking students. It was therefore decided that one session should focus on awareness-raising of critical pronunciation issues for French speakers of English and providing the learners with the tools they needed to overcome them. They are thus introduced to online pronouncing dictionaries like www.howjsay.com and www.youglish.com and shown how to use them, before checking the words they will need for a micro-speaking task. The concept of shadowing, whereby learners listen to and imitate the prosodic patterns of native speakers, is also explained to them as a way of improving prosody over time. We identified the key areas of difficulty where transfer from the L1 is likely. These included vowel sounds, word stress in transparent words (highly abundant in dental English), and prosody, which tends to be flat. To tackle these issues, we designed a series of scenarios that culminate in a micro speaking task, where students record their own productions, using their smartphones, to be sent to the teacher for personalised feedback during one of the tutorial sessions.

3.5. An example of Content and Language Integrated Learning (CLIL)

Another phase in the course is team-taught by the ESP teacher and dentistry teacher in a CLIL format; for more discussion of the varying degrees of CLIL see Taillefer (2013) and the Lanqua Project[7]. During this phase the dentistry teacher presents a clinical case to the students and engages in discussion with them. Although the students listen with interest to the case, they seem to pay

---

7. http://www.celelc.org/projects/Past_Projects/lanqua/index.html

little attention to how things are done linguistically. The ESP teacher's role is therefore to give more explicit guidance on how to produce the genre. As Hyland (2011) says, "ESP teachers cannot rely on subject specialists to teach disciplinary literacy skills as they generally have neither the expertise nor the desire to do so" (p. 9). However, the participation of the dentistry teachers in the ESP classroom also has the advantage of creating an English as a Lingua Franca (ELF) environment where the focus is on communicative competence. The students, who will be more likely to interact with other non-native speakers rather than native speakers in their professional practice, are thus in a more realistic and less threatening situation where clarity and effective communication take precedence over native-like competence. The students are motivated to see their dentistry teachers 'playing the game' and are possibly reassured to see that their own level of English can compare quite favourably with these researchers who participate in international conferences.

3.6. Tutorial sessions: individualised feedback

Individualised feedback is given in the tutorial sessions. For the first session, the students come with the materials they have gathered on their case (photos, x-rays) and describe it informally to their teacher. In the second tutorial, the students give a complete run-through of their presentation and are thus given feedback on structure, content, slides, pronunciation, delivery, accuracy of grammar and vocabulary, and overall clarity. As Whyte (2013) notes, "it is often the case that feedback on student performance in ESP and other language courses comes too late for reflection and improvement" (p. 15). It was therefore decided that feedback in the form of progress checks should be fully integrated into the course.

4. Assessment

For the final presentation, the examining panel is composed of the ESP teacher and a dental researcher, and the students deliver their presentation in front of their peers. The assessment criteria for the oral presentation are provided to the students in the penultimate class of the course. The criteria aim to take into account the multimodal nature of an oral presentation (Carter-Thomas & Rowley-Jolivet, 2003) by evaluating the accessibility and intelligibility of the English, the scope and accuracy of the grammar and vocabulary, the content and structure, the student's communication skills, the quality of their slides, and how well they deal with questions. Students are also assessed on how well they have applied the communication strategies highlighted by the course. The assessment grid[8] not only aims to harmonise grading practices among the eight different teachers on the examining panels, but is also designed to provide an itemised feedback report for the students on their performance. The criteria are broken down into five main areas, with each area earning the student a maximum of four points. The items reflect the objective of 'non-native fluency' rather than 'native likeness' (Pilkinton-Pihko, 2013). According to these criteria, students who give a B2 level performance or above and who fulfil all the content and communication criteria can achieve a mark of 20. Students in the audience are encouraged to ask questions after each presentation, with the incentive of earning 0.5-1 bonus points toward their own presentation grade. The presentations usually generate many questions from the peer audience as well as rich discussions with the dental professional on the examining panel. The latter's perspective is invaluable for assessing the disciplinary content, but it is the ESP specialist who focusses on the linguistic and communicative dimensions.

5. Course evaluation

At the end of the first class, the students completed an online questionnaire[9]. This pre-course questionnaire focussed in part on what Cheng (2011) has termed 'social milieu', that is to say the students' expectations of the course, their attitudes towards English, and the pertinence of the task, etc. Questions on their pre-course knowledge of clinical cases (structure, grammar, vocabulary, communicative strategies, and use of PowerPoint, etc.) would then allow us to

---

8. Supplement, part 1: https://research-publishing.box.com/s/adeudwll4uz7fh38g2qs1rjzmecqs08k

9. Supplement, part 2: https://research-publishing.box.com/s/adeudwll4uz7fh38g2qs1rjzmecqs08k

compare with the post-course responses as students are led to reflect on how they accomplished the task. The post-course questionnaire was completed at the end of the final presentations[10].

5.1. Analysis of student feedback

The feedback from the post-course questionnaire showed that in general the course was a positive experience for learners (Table 3). A majority of students had a positive opinion of the course and felt that they had succeeded in the final task, progressing both in English, in general, and, more especially, in scientific and professional communication.

Table 3. Post-course student feedback

	AGREE	DISAGREE	NEUTRAL	TOTAL
Positive course experience	48	4	10	62
ESP progress	47	8	7	62
EFL progress	35	13	12	60
Success in final task	49	6	6	61
Disciplinary boost	36	12	14	62
Tutorials helpful	50	7	5	62
Course materials helpful	46	6	10	62
Collaboration beneficial/ motivating	52	1	9	62

The students' responses also highlight the importance of the 'discourse domain', as we saw earlier. They felt that their disciplinary proficiency had helped them overcome difficulties in English and successfully complete the task. This was particularly apparent among the students who perceived themselves as 'weak' (Table 4). The weaker students felt their content expertise compensated for difficulties with English. However, almost half the group saw themselves as average and were divided equally in considering content knowledge only, or language and content skills combined as equally important in task success.

---

10. Supplement, part 2: https://research-publishing.box.com/s/adeudwll4uz7fh38g2qs1rjzmecqs08k

Almost half of the more proficient students also thought language and content strengths contributed equally. These students were also much more likely than other students to attribute their success to their language skills, with just under a third of them choosing this option. This was the only group to include students who thought language skills more important in their success, though a similar number chose the content option.

Table 4. Attribution of task success according to EFL proficiency

ATTRIBUTION of TASK SUCCESS	language only	%	language & content	%	content	%	TOTAL	%
strong students	6	29	10	45	5	24	21	34
average students	1	4	14	54	11	42	26	44
weak students	0	0	2	18	9	82	11	19
TOTAL	7	12	26	44	26	44	59	

It also appears that the course itself also helped the students overcome their difficulties. Table 5 clearly shows that the students felt more able to present a clinical case to their peers at the end of the course than at the beginning.

Table 5. Self-assessed skills before and after the course

	AGREE	DISAGREE	NEUTRAL	TOTAL	
Ability to present a clinical case	29	22	20	71	pre-course
	44	10	7	61	post-course
Knowledge of clinical case structure	29	24	19	72	pre-course
	56	2	4	62	post-course
Vocabulary proficiency	15	35	22	72	pre-course
	42	4	16	62	post-course
Phonology proficiency	12	44	16	72	pre-course
	35	13	14	62	post-course
Grammar proficiency	14	45	12	71	pre-course
	33	14	15	62	post-course
Communication skills	17	34	20	71	pre-course
	37	8	17	62	post-course
Confidence	17	45	10	72	pre-course
	29	23	10	62	post-course

The materials and tasks used appear to have contributed to this outcome, as we will now see.

At the outset, the students' knowledge of the structure of a clinical case presentation seemed sketchy and perhaps limited. However, by the end of the course, the vast majority felt they understood the organisational norms of this type of presentation, suggesting that our genre approach was effective. Similarly, they felt more able to use the appropriate lexical, grammatical, and to a lesser extent phonological features of this type of discourse at the end of the course than at the beginning. This improvement in their self-evaluated skills would seem to indicate that tasks such as mining corpora and pronunciation awareness activities do have a positive impact. The English course also seemed to be the context where communication skills were given special attention and students felt better equipped in these cross-disciplinary skills. The ELF classroom environment seemed to encourage communication and peer-to-peer interaction as the gains in confidence show. The CLIL element and our collaboration with our dentistry colleagues was also perceived in a positive light by the students (Table 3). This collaboration had a further beneficial effect for our dentistry colleagues as for some, it was a first step towards EMI (English Medium Instruction). Four of them were encouraged to follow the Teaching Academic Content in English (TACE) course run by the DLC for *Défi International* at Bordeaux University, as internationalisation becomes a key strategy for the university. However, overall, we are cautious not to over-extrapolate from these results given that our questionnaire was based on a seven-point Likert scale that could have induced a bias.

One particularly valuable aspect of the course, according to the students, was the tutorial sessions (Table 3). The students indicated that the progress checks and individual feedback provided in this setting played a key role in their performance, an opinion which was shared by the ESP teachers who taught the course. In summary, the combination of classroom instruction, personalised feedback, and semi-autonomous project work was coherent and effective for these students and their context. The efficacy of this blended format is in line with Hattie's (2008) meta-analysis:

"ideally, teaching and learning move from the task to the processes and understandings necessary to learn the task, and then to continuing beyond it to more challenging tasks and goals. This process results in higher confidence and greater investment of effort. This flow typically occurs as the student gains greater fluency and mastery" (p. 177).

6. Conclusion

If we refer back to the research questions that guided our study, it would seem, first of all, that the needs analysis helped in designing a course with academic and professional relevance, enabling students to draw on and communicate their disciplinary expertise. The programme from first year to fifth year is thus characterised by a shift in focus and input as students gain more disciplinary specialisation and play a more active role in their professional community.

Secondly, the post-course questionnaire indicates that the students had integrated the specific language features targeted in the activities and tasks. This, together with their disciplinary expertise, helped students successfully complete the task.

Finally, students perceived the individualised teacher feedback in the tutorial sessions to be highly beneficial. The development of this course has also led us to reflect on our role as ESP specialists, our domain of expertise, and how we can collaborate with subject specialists. From an institutional point of view, the collaboration with our dentistry colleagues has given us greater visibility and recognition within the faculty.

7. Acknowledgements

The materials for this course were developed and taught by a team of teachers at the *Département de Langues et Cultures*, Bordeaux University. We thank our colleagues, Valérie Braud, Anne-Laure Damongeot, and Thibault Marthouret for their valuable input throughout the design of this course.

References

Belcher, D. (2006). English for specific purposes: teaching to perceived needs and imagined futures in worlds of work, study, and everyday life. *TESOL quarterly, 40*(1), 133-156. https://doi.org/10.2307/40264514

Belcher, D. (2013). The future of ESP research: resources for access and choice. In B. Paltridge & S. Starfield (Eds), *The handbook of English for specific purposes* (pp. 535-551). Wiley-Blackwell.

Carter-Thomas, S., & Rowley-Jolivet, E. (2003). Analysing the scientific conference presentation (CP), a methodological overview of a multimodal genre. *ASp. la revue du GERAS, 39-40*, 59-72. https://doi.org/10.4000/asp.1295

Cheng, A. (2011). ESP classroom research: basic considerations and future research questions. In A. M. Johns, B. Paltridge & D. Belcher (Eds), *New directions in English for specific purposes research* (pp. 44-72). University of Michigan Press.

Dudley-Evans, T. (1993). Subject specificity in ESP: how much does the teacher need to know of the subject? *ASp. la revue du GERAS, 1*, 2-8. https://doi.org/10.4000/asp.4354

Dudley-Evans, T. (1997). Five questions for LSP teacher training. *Teacher education for LSP*, 58-67.

Dudley-Evans, T., & St John, M. J. (1998). *Developments in English for specific purposes: a multi-disciplinary approach.* Cambridge University Press.

European Commission. (2012). *First executive summary on language competences.* http://ec.europa.eu/dgs/education_culture/repository/languages/library/studies/executive-summary-eslc_en.pdf

Hattie, J. (2008). *Visible learning: a synthesis of over 800 meta-analyses relating to achievement.* Routledge.

Hyland, K. (2002). Specificity revisited: how far should we go now? *English for specific purposes, 21*(4), 385-395. https://doi.org/10.1016/S0889-4906(01)00028-X

Hyland, K. (2011). Disciplinary specificity: discourse, context and ESP. In A. M. Johns, B. Paltridge & D. Belcher (Eds), *New directions in English for specific purposes research* (pp. 6-24). University of Michigan Press.

Hyland, K. (2013). ESP and writing. In B. Paltridge & S. Starfield (Eds), *The handbook of English for specific purposes* (pp. 95-113). Wiley-Blackwell.

Hyland, K. (2015). Genre, discipline and identity. *Journal of English for Academic Purposes, 19*, 32-43. https://doi.org/10.1016/j.jeap.2015.02.005

Johns, A. M. (2013). The history of English for specific purposes research. In B. Paltridge & S. Starfield (Eds), *The handbook of English for specific purposes* (pp. 5-30). Wiley-Blackwell.

Ortega, L. (2015). Researching CLIL and TBLT interfaces. *System, 54*, 103-109. https://doi.org/10.1016/j.system.2015.09.002

Pilkinton-Pihko, D. (2013). *English-medium instruction: seeking assessment criteria for spoken professional English*. Dissertation. University of Helsinki.

Robinson, P. C. (1991). *ESP today: a practitioner's guide*. Prentice Hall.

Swales, J. (1990). *Genre analysis: English in academic and research settings*. Cambridge University Press.

Swales, J. M. (2011). Envoi. In A. M. Johns, B. Paltridge & D. Belcher (Eds), *New directions in English for specific purposes research* (pp. 271-274). University of Michigan Press.

Taillefer, G. (2013). CLIL in higher education: the (perfect?) crossroads of ESP and didactic reflection. *ASp. la revue du GERAS, 63*, 31-53. https://doi.org/10.4000/asp.3290

Van der Yeught, M. (2010). Editorial. *ASp. la revue du GERAS, 57*, 1-10.

Whyte, S. (2013). Teaching ESP: a task-based framework for French graduate courses. *ASp. la revue du GERAS, 63*, 5-30. https://doi.org/10.4000/asp.3280

Whyte, S. (2016). Who are the specialists? Teaching and learning specialised language in French educational contexts. *Recherche et pratiques pédagogiques en langues de spécialité, 35*(special 1). https://apliut.revues.org/5487

Wozniak, S., Braud, V., Sarré, C., & Millot, P. (2015). Pour une formation de tous les anglicistes à la langue de spécialité. *Les Langues Modernes, 3*, 67-76.

Wozniak, S., & Millot, P. (2016). La langue de spécialité en dispute. Quel objet de connaissance pour le secteur Lansad? *Recherche et pratiques pédagogiques en langues de spécialité. Cahiers de l'Apliut, 35*(spécial 1), 1-10.

Section 2.
Building confidence: addressing particular difficulties

5. Towards a dynamic approach to analysing student motivation in ESP courses

Daniel Schug[1] and Gwen Le Cor[2]

Abstract

This study seeks to understand student attitudes towards English for Specific Purposes (ESP) courses. Such courses were conceived, in part, under the belief that they would be inherently more motivating as they, ideally, correspond directly to students' interests and needs. Rather than accepting this notion at face value, this paper posits that a thorough analysis of student behaviours and attitudes in ESP courses is required to fully understand their effectiveness in terms of their capacity to motivate students. To do so, this paper suggests studying motivation through the lens of the Complex Dynamic Systems Theory (CDST). This theory allows for the analysis of motivation as a dynamic phenomenon, strongly dependent on all the factors present in a given system, namely, a language classroom. Questionnaires, interviews, and observation data were used to analyse student motivation in English for Arts and General English (GE) courses at a large, public university in France. Results indicate that students were mostly indifferent to the specialised elements of their language courses and that their motivation was more dependent on the structure of different activities.

Keywords: complex dynamic systems, L2 learning environment, motivation, English for specific purposes, L2 motivational self-system.

---

1. Università Ca' Foscari Venezia, Venice, Italy & Université Paris 8, TransCrit Research Lab, Paris, France; 956261@stud.unive.it

2. Université Paris 8, TransCrit Research Lab, Paris, France; gwen.le-cor@univ-paris8.fr

How to cite this chapter: Schug, D., & Le Cor, G. (2017). Towards a dynamic approach to analysing student motivation in ESP courses. In C. Sarré & S. Whyte (Eds), *New developments in ESP teaching and learning research* (pp. 73-91). Research-publishing.net. https://doi.org/10.14705/rpnet.2017.cssw2017.746

© 2017 Daniel Schug and Gwen Le Cor (CC BY)

Chapter 5

1. Introduction

Courses of ESP were conceived, in part, as the result of an interest in student learning motivation: these courses were thought to be inherently more motivating as they, ideally, correspond directly to students' interests and needs (Hutchinson & Waters, 1987). Despite some findings supporting this claim, Brown (2007) tells us that it has largely been accepted at face value with little evidence to back it up. Brunton (2009) reinforces this point with his argument that, when given a choice, students often prefer GE courses to ESP ones.

Therefore, the objective of this study is to determine precisely what elements affect student motivation in ESP courses. This paper relies principally on CDST; this theory, combined with other theoretical constructs, allows for a thorough analysis of a learner's attitudes at a given moment and an understanding of how they are influenced by numerous factors in and out of the learning environment (Larsen-Freeman, 2015). After summarising some major developments in motivation research, we examine a small sampling of studies conducted under the CDST and how their findings apply to the ESP context. Research methods are presented with data from a study conducted in ESP courses at a French university; the results are compared with GE courses to determine how student attitudes differ.

2. Review of literature

Gardner's (1960) Socio-educational Model (SM) has long dominated studies in L2 learning motivation in various contexts, including ESP (Ushioda, 1996). Its main tenet is integrativeness; a strong integrative orientation comes from a desire to identify with people of the L2 community and often leads to a strong learning motivation. The model also includes an instrumental orientation, occurring when one studies a language for professional gain or due to another external force (Gardner, 1960).

Applying the SM to ESP courses at a Yemeni university, Al-Tamimi and Shuib (2009) used questionnaires and interviews to classify student motivation.

Participants reported strong instrumental orientations, such as learning English for their future jobs, with many reporting negative feelings towards the L2 community. While this study provides useful insights on the general attitudes of these students, researchers have called for more thorough analyses of the learning environment, given its capacity to cause frequent motivational changes during a lesson (Bier, 2013; Lavinal, Décuré, & Blois, 2006).

Responding to this call, Dörnyei (2009) devised the L2 Motivational Self-System (L2MSS) to focus more directly on a learner's attitudes towards the classroom environment. The model contains several constructs, but this study will focus on how motivation is affected by the L2 Learning Environment (L2LE): this construct includes relationships with one's peers and teachers, the layout of the classroom, the nature of the learning activities, and innumerable other factors that can affect students' attitudes towards learning. To understand how L2LE elements work with other factors and translate into classroom behaviours, this study draws from past research using the CDST; this theory represents an effective framework for analysing the motivating elements present in ESP courses, given its insistence on classroom experiences and real-time observations (Henry, 2015; Waninge, 2015).

While studying high school students in a foreign language classroom, Waninge (2015) asked students about the emotions they felt during their lessons. Among the most cited emotional states were interest (leading to active participation in the classroom), boredom (leading to disengagement), and neutral attention (leading to passive listening). With many participants, Waninge found that interest was the result of contextual factors, such as learning activities, the teacher, and peers. Interest was also sparked when activities related to students' personal interests and pre-existing non-language goals. While this latter finding may point to the potential for ESP to stimulate student interest, the former most definitely shows the need to consider all factors in the classroom when studying student engagement.

Other CDST studies reinforce the need to analyse all elements in a classroom and point to the fact that course content and activities are only part of the countless factors affecting motivation. Such conclusions indicate that future research should depend more on class observations to have a more detailed view of participant behaviours (Henry, 2015).

Building on the findings of these CDST experiments, the present study includes classroom observations that focus directly on student participants, in addition to interviews and questionnaires, in both ESP and GE courses, to answer the following research questions:

- What elements do students find motivating in their language courses?

- How do these elements differ between ESP and GE courses (using English for Arts as a case study)?

3. Methods

This study was conducted at Paris 8 University in France during the 2016/2017 academic year, with students in English for Arts courses at A2 and B1 levels. These courses are designed based on the teacher's area of expertise and experience, sometimes in consultation with the Arts department. The levels are based on the Council of Europe's (2001) framework for language proficiency. To provide a basis for comparison, motivation data from students in GE courses were also included.

In order to increase the validity of our results (Dörnyei, 2011), we used a mixed methods approach – via questionnaires, interviews, and observation schemes – for collecting data on student motivation and attitudes in their English classes.

3.1. Questionnaire

The motivation questionnaire we used was adapted from previous L2MSS studies (Brander, 2013; Tort Calvo, 2015; You & Dörnyei, 2016) to create a tool for studying learner attitudes towards the learning environment. The final version of the questionnaire contained 27 items. As this data is part of a larger study, only answers from the six questions measuring the L2LE are reported here[3]. A seventh item was eliminated from consideration due to an anomaly in the printed version. The questionnaire was given to students in French and was distributed to some via email and to others in paper format. In total, it was sent to 42 GE A2 students, 64 GE B1 students, 55 ESP A2 students, and 61 ESP B1 students. Their participation was optional.

It was deemed necessary to provide a shorter questionnaire than was used in past studies, given a low response rate of about 50% from the pilot session. Nonetheless, special care was given to assure a fair representation of the different constructs.

3.2. Class observations

All class observations were conducted by the first author of this paper, in courses of ESP and GE at the A2 and B1 levels. An adapted version of the observation scheme used by Guilloteaux and Dörnyei (2008) was used to track motivation during the language courses, operationalised as visible student engagement[4]. Observing learning engagement is a common approach to studying classroom or task-based motivation, as student participation in learning tasks is indicative of the willingness and persistence often associated with motivated students (Guilloteaux, 2007).

In each class, one student was chosen as the focus of the observations to understand behaviours in a more in-depth way (Henry, 2015). A convenience sampling strategy was deemed most appropriate for identifying participants, given the importance of interviewing students immediately following a lesson (Henry, 2015). First, students had an opportunity to say they did not wish to participate in interviews or observations. Next, the researcher asked students who would not be available for an interview directly after the lesson; these students

---

3. See 'The questionnaire': https://research-publishing.box.com/s/w2tk0oe7royle3qw90ceoueu0c9nswer

4. See 'The observation scheme': https://research-publishing.box.com/s/w2tk0oe7royle3qw90ceoueu0c9nswer

were immediately eliminated from consideration. Finally, as the researcher needed to be in a discreet position within the classroom, only students who were easily visible from such a position could be considered. Ultimately, these criteria generally left only two or three possible participants in each group, from which one was chosen randomly.

The student was observed throughout six three-hour lessons, which is the typical duration of non-specialist language courses at this university. Behaviours were then coded into three categories: passive engagement (A), active oral or written participation (P), and disengagement (D). Category A was noted when students reacted to stimuli, kept their eyes on the speaker, or copied information without adding anything original. P was noted when students participated in group tasks, actively completed individual work, and raised their hands, adding to class discussion. Finally, D was noted when the student was not paying attention, evidenced by checking their phone, sleeping, chatting with classmates about personal matters, or staring away from the class. A last category (N) was also created for non-academic time, to account for late starts or early endings to the lesson.

3.3. Interviews

Each observation participant completed two interviews with the first author, one in the first half of the semester, and another towards the end; semi-structured interviews were used to assure that certain points were addressed while still allowing students to express unexpected opinions (Drever, 2003). Questions asked students to comment on the emotions they felt during their language lessons and how their motivation fluctuated over the course of the three hours and over the whole semester; students were asked to illustrate these fluctuations with line graphs. Additional questions gathered information about students' past experiences learning English and how they planned to use the language in the future (Henry, 2015). An indicative list of the principal guiding questions can be found in supplement materials[5].

---

5. https://research-publishing.box.com/s/w2tk0oe7royle3qw90ceoueu0c9nswer

As several participants expressed discomfort with the interviews being recorded, notes were taken by the researcher, then interview data were emailed to participants. Students verified the accuracy of their answers and could modify them as necessary; this was done within twenty-four hours of the interview.

4. Results

The following tables contain a summary of the results of this study, focussing on class behaviours. As stated, the interview and observation data come from one student from each of the courses observed (one ESP Arts A2, one ESP Arts B1, one GE A2, one GE B1). Table 1 below represents the number of respondents for each questionnaire.

Table 1. Number of questionnaire respondents per group

Course	ESP Arts A2	ESP Arts B1	GE A2	GE B1
Number of Questionnaires Distributed	55	61	42	64
Number of Respondents	38	45	30	46
Response Rate (rounded to the nearest percent)	69%	74%	71%	72%

4.1. Questionnaire data

Table 2 shows the proportion of students' positive reactions to different aspects of the classroom environment. For example, the figure regarding teacher/student relationship comes from the number of respondents marking 4 or 5 on the Likert scale for that item. For True/False questions, the figure is based on the number of respondents answering True. These results tend to show that students in all groups have similar attitudes towards their courses, with most being positive. Still, a majority do not feel a sense of excitement when going to class, and large numbers admit to giving into distractions. Aside from these measures, it seems many students feel their class is useful, practical, and that the teacher creates a good classroom atmosphere.

Table 2. Students' positive reactions to the learning environment

Course	ESP A2	ESP B1	GE A2	GE B1
The teacher/student have a good dynamic	92.1%	100%	93.3%	91.3%
I'm excited about going to class every week	34.2%	15.5%	16.6%	26.1%
I can ignore other distractions during lessons	52.6%	40%	66.7%	60.9%
The course is useful and practical	92.1%	82.2%	96.7%	89%
The course corresponds to my level	84.2%	71.1%	93.3%	84.8%

The questionnaire also contained one open-ended question related to the learning environment, asking students to comment on an activity they found particularly useful. This item was the only one without a 100% response rate. The answers to this question are as follows.

In ESP A2, the response rate was 78.9%, (30 answers out of 38 participants). Three respondents (=10%) referenced specialised materials. All other responses described activities that helped develop general language skills; 16 students appreciated oral activities, four mentioned listening comprehension, and others covered skills such as pronunciation and grammar.

In ESP B1, the response rate was 51.1% (23 answers out of 45 participants). Of these responses, three (=13%) contained references to specialised materials. All other responses described activities related to general language skills; 13 talked about oral skills, two talked about general cultural knowledge, and three referenced listening comprehension.

Of the responses in GE A2 (63.3% response rate, i.e. 19/30), most students talked about general language skills; five focused on oral skills, five talked about grammar, six referenced listening and reading comprehension. The remaining responses discussed expanded cultural knowledge.

Finally, in GE B1 (76% response rate, i.e. 35 answers out of 46 participants) most students talked about general language skills; 18 students talked about oral activities and presentations, six talked about listening and reading comprehension, and five referenced grammar activities. A variety of other responses were present, including general compliments of the teacher as well as increased cultural awareness.

Overall, very little difference can be seen between groups: students from all four groups reported their appreciation of similar activities (based on the development of general language skills and on cultural knowledge) as well as the benefits of such activities. Very few students (11.5% in total) in the ESP groups considered specialised materials useful and, therefore, potentially motivating.

4.2. Interview data

The following figures present interview data, with students' line graph representations of motivational changes over the course of one lesson, in either Week 3 or 4 of the course, and over the course of the semester.

Figure 1. ESP A2 Student 1's motivational change over the course of a three-hour lesson (3pm-6pm, solid line) and the semester (September-December, dotted line)

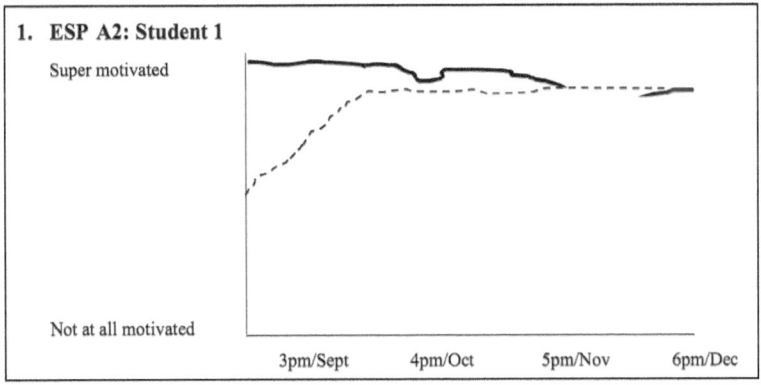

In Figure 1, Student 1 reported entering the lesson with a high level of motivation, as is her habit in most of her courses. This motivation fell slightly, however, when the teacher had them review an activity they had done the previous week, as she felt frustrated at spending more time on an already completed task. This resulted in a small motivation decrease throughout the three-hour lesson, as the student reported the same complaint about other activities that day. Still, her attitude remained relatively positive and she reported feelings of "joy" and interest throughout the lesson, originating from her relationship with her peers and all the new information she was learning. She described only rare moments of disengagement, characterised by doodling in her notebook or checking her phone; these were more out of habit than a problem with the lesson.

Regarding her semester-long motivation, she indicated starting the semester with a relatively negative outlook for the course, as she had never liked English classes and only took this one out of obligation. This attitude changed, however, as illustrated by the steady increase in her reported motivation between September and October, because the activities sparked an interest for her and she appreciated that the course was specialised for art students.

Figure 2. ESP B1 Student 2's motivational change over the course of a three-hour lesson (12pm-3pm, solid line) and the semester (September-December, dotted line)

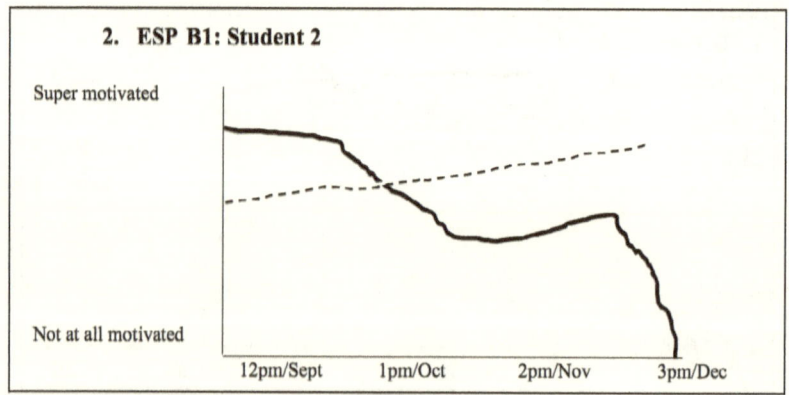

As shown in Figure 2, Student 2 reported being very motivated at the beginning of the lesson, because she always loved English and because the cultural lesson in the beginning was very interesting. The motivation fell, however, largely because of the duration of the course; the student reported being too tired and too hungry to pay attention. Moreover, she claimed a reading activity in the second portion of the lesson was too easy to be motivating. The initial interest was therefore replaced by fatigue and lassitude. Regarding moments of complete disengagement, in which the student was on her phone or chatting, the student said that it happens in all her classes when she must listen to a long speech or do the same activity over a long period.

Figure 2 also shows that Student 2 started the semester with a medium level of motivation. Though she did appreciate being in an ESP course, she claimed the determining factor in her steadily increasing motivation over the semester was her classmates; as the semester went on, they got to know each other better and enjoyed doing group work together.

Figure 3. GE A2 Student 3's motivational change over the course of a three-hour lesson (12pm-3pm, solid line) and the semester (September-December, dotted line)

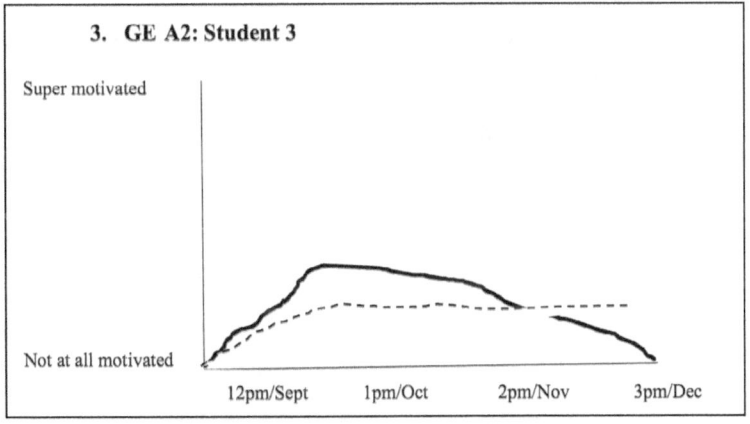

Regarding Figure 3 and Student 3's attitudes over the three-hour lesson, this student reported always entering all her courses with low motivation. When class starts and she has a task to complete, motivation increases. The increase was only temporary, however, because she struggled to stay engaged over three hours.

The high point was the result of group work, which she enjoys; this was the only time she reported a positive emotion of interest. During the rest of the lesson, she reported fatigue, from having to work for three hours, boredom from spending too much time on difficult tasks, and indifference, for grammar and vocabulary activities. Still, she does her best to never fully disengage; she uses her phone or chats with friends only when activities become too difficult or too dull.

Figure 3 also indicates that Student 3 started the semester with a negative attitude; she described never liking her English courses in the past and having teachers that failed to spark her interest. This course, however, did motivate her a little, because the professor gave students a very active role and presented materials that were necessary for general, cultural knowledge. Her motivation will never be very high, she explained, as she feels she will never use English in her daily or professional life.

Figure 4. GE B1 Student 4's motivational change over the course of a three-hour lesson (3pm-6pm, solid line) and the semester (September-December, dotted line)

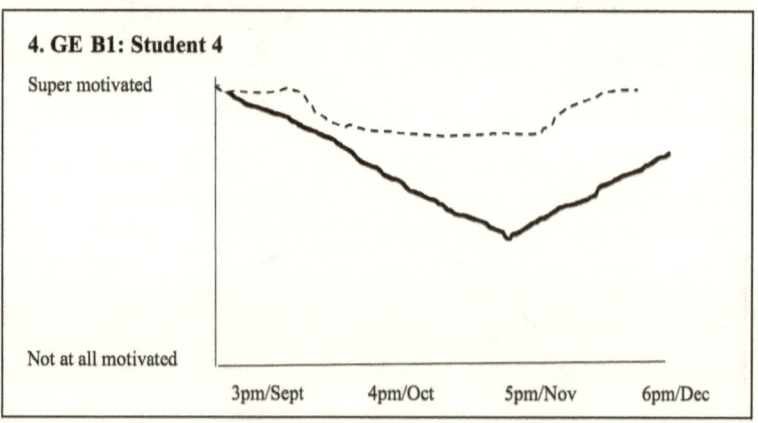

As can be seen in Figure 4, Student 4 entered the classroom highly motivated, because he has always loved his English courses; he is also particularly engaged whenever he can participate in a discussion. He initially lost motivation on this day, even to the point that he stopped working, because he was sick and tired. Ultimately, he regained motivation at the end to complete a translation activity, as he knew it would be important for an exam. He described that this day was rather exceptional, given his illness; he described feeling ill during the lesson and unable to concentrate. He still referenced interest, however, because he was fascinated by the gender studies activities presented in this lesson. Otherwise, moments of complete disengagement were normally out of habit, rather than an actual problem with the course.

Student 4 also indicated being highly motivated all semester, but claimed that it was difficult in October because he was busy with other courses and obligations and could not focus on English work. In November, after failing the midterm, he reported anger and discouragement. Moving forward, however, his interest in class activities was enough to re-motivate him through December.

4.3. Observation data

Table 3 below contains percentages showing how often each student exhibited each type of motivation over the course of 18 hours of observation, made up of six three-hour observation sessions. It also shows how often students changed between active and passive engagement and disengagement, and the average number of changes per hour.

In ESP A2, Student 1's interview answers (reported in the previous section) reinforce her observed behaviours shown in Table 3. Six instances of disengagement were observed, of which four corresponded to moments in which students spent extended periods on one activity; in each of these cases, the activity was a specialised activity, relating to art. The other two instances occurred at the beginning of the class, as the teacher presented the day's lesson; these instances are likely explained by her comment about sometimes texting out of habit. Instances of passive engagement occurred with highest regularity; these seemed to correspond to the teacher's instructions. When the students were asked to participate in a small group discussion, Student 1 exhibited active engagement; when they were asked to read, follow a video or participate in

Chapter 5

whole-class discussions, she was passive. These observations appear to be true regardless of the activity, whether specialised to art students, general cultural presentations, or grammar activities.

Table 3. Students' observed motivational distribution over the semester. Note that figures are rounded to the nearest whole percentage, meaning some students' data may add up to slightly more than 100%

Student	Active Engagement (P)	Passive Engagement (A)	Disengagement (D)	Non-Academic Time (N)	Number of observed motivational changes	Average number of changes per hour
Student 1 (ESP A2)	31%	50%	2%	17%	81	3.7
Student 2 (ESP B1)	18%	54%	11%	18%	99	4.6
Student 3 (GE A2)	24%	45%	14%	18%	89	3.6
Student 4 (GE B1)	33%	37%	9%	21%	90	4.7

In ESP B1, the observations presented in Table 3 also echo what Student 2 said in her interview. She had 34 observed instances of disengagement, accounting for more than 10% of the 18 hours of observation; of these cases, 29 seemed to have been triggered by completing an activity she deemed too passive, such as reading or listening followed by answering questions, or by spending too much time on an activity. Of these 29 instances, 20 occurred during ESP activities, such as discussing a work of art or reading about an art movement. The interest she expressed in cultural lessons carried over to other lessons as well, resulting in long periods of passive attention, some lasting up to 30 minutes, when the professor presented the cultural background of works of art to be studied. Instances of active engagement, though numerous, were often short-lived. While they did get increasingly longer later in the semester (some periods of active attention lasting nearly 25 minutes in December, as opposed to short ones around nine minutes on average in September), no instances of volunteering information were observed; Student 2 only seemed to participate when in groups or specifically asked by the professor.

Student 3 (from the GE A2 group) exhibited the highest levels of observed disengagement of the four participants, as illustrated in Table 3. The fatigue she mentioned in the interviews seemed to be a major factor in her lessons, as her disengagement was often marked by sleeping in class. The difficulty of the task also seems to play a role; ten of the 15 observed instances of disengagement occurred during speaking or listening activities. These numbers are in stark contrast with the high levels of uninterrupted passive engagement, with some instances as long as 45 minutes, relating to correcting pre-prepared work in class, suggesting Student 3's interest in less spontaneous tasks.

Finally, in GE B1, Student 4's motivation fluctuated considerably in the classroom throughout the semester. His self-reported motivation illustrated in Figure 4 appears to be an accurate representation of his attitudes during the course, as only two instances of disengagement were observed in September; one in December, and the remaining 17 during the lessons observed in October and November. Student 4's comment about being excited when participating in discussions also rang true; his 35 instances of active engagement almost exclusively happened during moments of class discussion and group work. Indeed, the only times he was passive in class was when someone else was talking.

5. Discussion

The present study seeks to shed light on motivating elements in specialised language courses to better understand their motivational value in relation to GE courses. A combination of questionnaires, interviews, and classroom observations helped to identify some elements of a course that students found motivating. Considered together, the above results support Henry's (2015) insistence on considering numerous factors when analysing student motivation and validates the use of CDST. Students reacted to the duration of the class, work with their peers, the timing of class activities, and the structure of different tasks.

As mentioned previously, questionnaire data showed that students in all four groups had mostly positive attitudes towards their courses (considering them useful and practical), even though most admitted not being excited about going to class, as well as being easily distracted during a lesson. In the open-ended question, students in all groups seemed to appreciate similar activities, focusing on general language skills and cultural knowledge. Aside from the six ESP students who mentioned specialised materials (11.5% of responses in the ESP groups), all respondents described benefitting from essentially the same type of activities. Such a finding fits in with past research comparing students in ESP and GE courses, where students report a preference for GE work, as it allows them to develop L2 skills that they could use in a wider range of contexts (Brunton, 2009).

Students' responses in the interviews seem to reinforce what was found in the questionnaire; ESP only received minimal attention in the interviews, and only from the ESP A2 student: she appreciated the change ESP offered from her past learning experiences. Such a point perhaps ties in to the students' attitudes against repeating or dwelling on tasks. As Terrier and Maury (2015) describe, most students in France should finish high school with a minimum English level of B2, as this is the target level of the language classes offered at this level of education (and in the *Baccalauréat* exam): thus, having to go over the same A2 grammar points in a GE course at university could be viewed as a demotivating factor. Therefore, the novelty of ESP courses perhaps provided a welcome change.

Moreover, similar learning activities, such oral activities, comprehension activities, and activities that expand their cultural knowledge, were found to be motivating in all groups, eliciting both active and passive engagement. Passive engagement was strong in all groups, as in Waninge's (2015) study, often occurring during lectures, classmate presentations, or listening activities. Instances of active engagement were most often the result of group work, reinforcing Henry's (2015) point regarding the influence peers could have on motivation. These findings suggest that learning activities have a strong influence on the type of engagement students exhibit; their responses and behaviours show that they associate certain behaviours with certain learning activities. For example, classmate presentations could only stimulate passive engagement, while group work provides an atmosphere more conducive to active participation. Further studies may wish to analyse this relationship in greater detail.

Also, similar to Henry's (2015) study, students in both ESP and GE courses reported being demotivated by difficult tasks and by spending too much time on one activity. While the tasks that are considered difficult vary from student to student, the maximum amount of time participants seem able to spend on an activity is around fifteen minutes, after which they tend to disengage.

6. Conclusions

This study lends some support to Brown's (2007) claim that ESP's alleged motivating element is a "folk assumption". Aside from the Student 1's' claim that ESP presented a welcome change, and the fact that only 11.5% of respondents in the ESP groups mentioned ESP materials as being interesting and useful (and therefore potentially motivating), all participants in GE and ESP groups seemed to appreciate the same elements of their courses; this finding indicates that students are at least partly indifferent to specialised activities. Oral and group tasks, for example, seemed to be motivating, regardless of whether the task was specialised. While further work is needed in this domain, this study suggests that the classroom environment and learning activities may have a stronger importance than specialised course content, at least with regard to Arts students.

Some of the potential limitations of this study relate to the sample of four observed students. Aside from being a very small sample, the convenience sampling strategy used here could not ensure the representativeness of the participants. While this method was chosen following Henry's (2015) advice to focus more on individual students and conducting interviews directly after the lesson, additional studies should be performed using larger and more representative samples. Additionally, while the 12-week semester is typical of French universities, it would be interesting to spend time in a context in which students stay in the same learning environment for a longer period to have a better idea of how minute-to-minute fluctuations provoke long-term changes. Similarly, fluctuations in motivation could also be investigated in shorter ESP classes (1.5 to 2 hour classes) in order to consider the impact of class duration on student motivation. Finally, it might be worthwhile to conduct a similar study in a context where ESP courses are designed following systematic student needs' analyses, so that results align more closely with standard ESP research.

References

Al-Tamimi, A., & Shuib, M. (2009). Motivation and attitudes towards learning English: a study of petroleum engineering undergraduates at Hadhramout University of Sciences and Technology. *GEMA: Online Journal of Language Studies*, *9*(2), 29-55.

Bier, A. (2013). The motivation of adolescent pupils to learn English as a foreign language: a case study. *EL.LE 2*(2), 429-459. https://doi.org/10.14277/2280-6792/62p

Brander, A. (2013). *Developing language learners with Dörnyei: a study of learning environments and motivation at a Swedish upper-secondary school* (Unpublished bachelor thesis). Halmstad University, Halmstad, Sweden.

Brown, D. (2007). Language learner motivation and the role of choice in ESP listening engagement. *ASp, 51*, 159-177. https://doi.org/10.4000/asp.579

Brunton, M. (2009). An evaluation of students' attitudes to the general English and specific components of their course: a case study of hotel employees in Chiang Mai, Thailand. *ESP World, 4*. http://esp-world.info/Articles_25/ESP%20world%20study.pdf

Council of Europe. (2001). *Common European framework of reference for languages: learning, teaching, assessment*. Cambridge University Press.

Dörnyei, Z. (2009). The L2 motivational self-system. In Z. Dörnyei & E. Ushioda (Eds), *Motivation, language identity, and the L2 self* (pp. 9-42). Multilingual Matters.

Dörnyei, Z. (2011). *Research methods in applied linguistics: quantitative, qualitative, and mixed methodologies*. Oxford University Press.

Drever, E. (2003). *Using semi-structured interviews in small-scale research: a teacher's guide*. Scottish Council for Research in Education.

Gardner, R. (1960). *Motivational variables in second language acquisition* (Unpublished doctoral dissertation). McGill University, Montreal, Canada.

Guilloteaux, M. J. (2007). *Motivating language learners: a classroom oriented investigation of teachers' motivational practices and students' motivation* (Unpublished doctoral dissertation). University of Nottingham, Nottingham, UK. https://doi.org/10.1002/j.1545-7249.2008.tb00207.x

Guilloteaux, M. J., & Dörnyei, Z. (2008). Motivating language learners: a classroom-oriented investigation of the effects of motivational strategies on student motivation. *TESOL Quarterly, 42*(1), 55-77.

Henry, A. (2015). The dynamics of possible selves. In Z. Dörnyei, P. MacIntyre & A. Henry (Eds), *Motivation dynamics in language learning* (pp. 83-94). Multilingual Matters.

Hutchinson, T., & Waters, A. (1987). *English for specific purposes: a learning-centered approach*. Cambridge University Press. https://doi.org/10.1017/CBO9780511733031

Larsen-Freeman, D. (2015). Ten 'lessons' from complex dynamic systems theory: what is on offer? In Z. Dörnyei, P. MacIntyre & A. Henry (Eds), *Motivation dynamics in language learning* (pp. 11-19). Multilingual Matters.

Lavinal, F., Décuré, N., & Blois, A. (2006). Impact d'une première année d'IUT sur la motivation des étudiant/es à apprendre l'anglais. In *Cahiers de l'APLIUT*[En ligne], *25*(1). https://doi.org/10.4000/apliut.2599

Terrier, L., & Maury, C. (2015). De la gestion des masses à une offre de formation individualisée en anglais-LANSAD : Tensions et structuration. *Recherche et pratiques pédagogiques en langues de spécialité, 34*(1), 67-89. https://doi.org/10.4000/apliut.5029

Tort Calvo, E. (2015). *Language learning motivation: the L2 motivational self system and its relationship with learning achievement* (Unpublished bachelor thesis). Universitat Autònoma de Barcelona, Barcelona, Spain.

Ushioda, E. (1996). *The role of motivation. Learner autonomy.* Authentik Language Learning Resources Ltd.

Waninge, F. (2015). Motivation, emotion, and cognition: attractor states in the classroom. In Z. Dörnyei, P. MacIntyre & A. Henry (Eds), *Motivation dynamics in language learning* (pp. 195-213). Multilingual Matters.

You, C. J., & Dörnyei, Z. (2016). Language learning motivation in China: results of a large-scale stratified survey. *Applied linguistics, 37*(4), 495-516. https://doi.org/10.1093/applin/amu046

6 Teaching compound nouns in ESP: insights from cognitive semantics

Marie-Hélène Fries[1]

Abstract

The objective of this chapter is to explore the relevance of cognitive linguistics for teaching [noun] + [noun] constructions to French learners of English for Specific Purposes (ESP), and more specifically, for process engineering. After a review of research on Compound Nouns (CNs) and explicit versus implicit learning, three basic tenets of cognitive linguistics are highlighted: the encyclopedic meaning of words (i.e. drawing on specialised knowledge in order to understand [noun] + [noun] compounds), the continuum between grammar and lexicon (learning CNs as terms, rather than as a grammar rule), and the symbolic nature of language (explaining CNs graphically). This chapter then reports on a case study of the use of [noun] + [noun] constructions by French learners of English for process engineering. Learners received explicit instruction on compound noun formation in two conditions: the experimental group (eight students) were taught via a cognitive-semantic approach, while the control group (eight students) received training in grammar (morphosyntactic approach). Data include CNs produced by the learners in summaries based on note-taking from specialised videos and in slideshows for internship presentations. Analysis and discussion claim an advantage for the experimental group in terms of implicit learning of specialised CNs, stylistic accuracy, and relevance of the graphical representation of CNs.

Keywords: cognitive linguistics, compensation strategies, compound nouns, implicit learning, specialised domains.

---

1. Université Grenoble-Alpes, Grenoble, France; marie-helene.fries@univ-grenoble-alpes.fr

How to cite this chapter: Fries, M.-H. (2017). Teaching compound nouns in ESP: insights from cognitive semantics. In C. Sarré & S. Whyte (Eds), *New developments in ESP teaching and learning research* (pp. 93-107). Research-publishing.net. https://doi.org/10.14705/rpnet.2017.cssw2017.747

1. Introduction

Noun Phrases (NPs) cover a wide grammatical category which can include adjectives (as in 'high pressure liquid chromatography'), participles (for instance 'scanning tunnelling microscope'), adjunct nouns (as in 'computer software'), and clauses, etc. This study has been prompted by research showing that NPs are far more frequent in English texts for Science and Technology (S&T) than in general English (Biber & Gray, 2016; Salager-Meyer, 1984), corroborated by the repeated observation, in the author's teaching experience, that [noun] + [noun] constructions are especially difficult to master for French S&T students.

Although CNs have officially been part of the English syllabus for undergraduate S&T students in the University of Grenoble since 1990 at least (Upjohn, Blattes, & Jans, 2013) and are now formally taught again in some Master's degree classes, especially chemistry and process engineering, students seem to have problems with the [noun] + [noun] construction, especially those who have a weaker level of English (A2 level on the Common European Framework of Reference for languages (CEFR) especially, and B1 in some measures). These difficulties can be largely explained by a comparison between French and English NPs. French is a Romance language and favours post-modification of the head noun, placing it first (for example technologie de l'information), whereas English, as a Germanic language influenced by Latin, mostly allows pre-modification, with the head noun coming last (information technology). In the British National Corpus, for example, Rossi, Frérot, and Falaise (2016, p. 175) counted about 12 times as many CNs as 'of' constructions per million words (16,460 [noun] + [noun] compounds versus only 1,330 instances for [noun] of [noun] NPs). In addition, whereas the relationships between the different nouns are stated explicitly in post-modification, through the use of prepositions and determiners, these semantic links are absent in pre-modification, as the [noun] + [noun] construction is simply based on juxtaposition (Downing, 1977). This may result in ambiguity in the absence of a clear context. For example, in S&T, a 'satellite company' can either be a type of subsidiary (metaphorically, in economics) or a firm producing and/or selling satellites.

The [noun] + [noun] construction is especially important in S&T, because it is a common way of coining new terms (Cabre, 1999). This issue is even crucial in some disciplines, such as chemistry, where the International Union of Pure and Applied Chemistry favours the [noun] + [noun] construction in naming chemical compounds (carbon dioxide, sodium chloride, etc.). The difficulties French students encounter with CNs are therefore professional, terminological and grammatical issues, which can best be addressed at the interface between ESP and didactics. The meanings of the word 'compound' seem to exemplify this link: in grammar, 'compound nouns' means nouns consisting of two or more words (and more particularly nouns in Upjohn et al., 2013), whereas in chemistry, a compound is a substance composed of two or more atoms.

The present study will start with a review of research already conducted on CNs in linguistics and terminology, as well as with the distinction between explicit and implicit learning in Second Language Acquisition (SLA), a dichotomy which could be useful in order to take a step back from the local teaching context. The relevance of a cognitive-semantic approach to CNs will then be explored through a case study conducted with 16 lower-intermediate Master students in process engineering during autumn 2016.

2. Research framework

Research on NPs in English includes various linguistic approaches, among which generative theory (Lees, 1963; Levi, 1978), pragmatics (Bauer, 1979), cognitive views (Benczes, 2011), and corpus studies (Bauer & Renouf, 2001; Biber & Gray, 2016). It covers morphology and syntax (Berent & Pinker, 2007; Olsen, 2000), phonology (Fudge, 1984; Giegerich, 2004), and semantics (Ryder, 1994). It is applied in translation studies (Maniez, 2007; Torres, 2015) as well as language acquisition (Parkinson, 2015; Wilches Alvear, 2016). Corpus studies are especially relevant for ESP, because they provide evidence for the frequency of written [noun] + [noun] compounds in science. Biber and Gray (2016, p. 148), for example, have found that the number of pre-modifying nouns increased steeply during the 20th century in their S&T sub-corpus.

Morphology, syntax, semantics and phonology are precious resources for English teachers, enabling them to understand the compounding process better. SLA studies have shown that learners' first languages (L1) have an influence on their use of [noun] + [noun] compounds. Parkinson (2015), for example, found that native speakers of Mandarin Chinese, whose L1 allows [noun] + [noun] constructions, use compound nouns more correctly in their writing than learners whose L1 does not allow such constructions, such as Portuguese students. Moreover, because [noun] + [noun] compounds often become lexicalised, students routinely encounter them every time they come across texts or videos linked to their main field of study, which implies that they could be learnt implicitly (through exposure to specialised language), as well as explicitly (thanks to the teaching of syntax and practice exercises, which are still common in France). The concept of implicit learning can be traced back to Krashen (1981, 1987) and has been the focus of two thematic issues of the journal Studies in Second Language Acquisition (vol. 27(2) 2005, vol. 37(2) 2015). There is now widespread agreement that explicit and implicit learning can be seen as two autonomous but interrelated processes (Ellis, 2003, 2005; Hulstijn, 2005). This quick survey of the research on [noun] + [noun] compounds suggests that it stands at the crossroads between ESP and didactics, drawing its basic materials (terms, texts and concepts) from specialised domains, and its epistemological stance (grammatical versus lexical, explicit versus implicit) from a conceptualisation of language learning.

At the University of Grenoble, the current English course in the Masters in Process Engineering allows for one hour of explicit teaching of CNs (including practice exercises). It would be difficult to devote more teaching time to the [noun] + [noun] constructions, as contact hours are limited (36 hours in the first year and 24 hours in the second) and all the Master students in process engineering need to validate the three competences the Association of Language Teachers in Europe considers essential for students at the B2 level at the crossroads of linguistic and professional skills, namely to be able to "give a clear presentation on a familiar topic, [...] scan texts for relevant information, [...] and make simple notes that will be of reasonable use for essay or revision purposes" (Council of Europe, 2001, Appendix D, pp. 251-257). Accordingly, the syllabus focuses on Task-Based Language Teaching (TBLT): poster sessions based on popular science articles or research papers (to improve reading comprehension) for the first year and oral presentations and note-taking (based on process engineering topics), in the second year.

In these conditions, it seems tempting to explore alternative ways of explaining CNs that encourage implicit learning. Cognitive semantics seem to offer a suitable theoretical framework in this perspective. Firstly, cognitive linguists posit that the meaning of words is encyclopaedic and includes everything a user knows about these words (Croft, 2003; Langacker, 1987). This prompts a first research question: does the encyclopaedic meaning of words encourage S&T students to use the knowledge they have of their specialised domains to understand CNs? In other words, when French S&T students are in a situation where they are not just required to learn new words in order to please their English teacher, but really need to use the terminology of their specialised field while completing a 'real life' task, does it encourage implicit learning of [noun] + [noun] constructions? Secondly, cognitive linguistics has shown that there is a continuum between grammar and lexicon (Langacker, 1987, p. 3). This is exemplified by many lexicalised [noun] + [noun] compounds, which are examples of composition in the nominal group but have also found their way into dictionaries (e.g. iron oxide). This leads us to a second research question: could repeated exposure to key lexicalised CNs in a given specialised domain help students to use these CNs correctly? Thirdly, cognitive semantics, while agreeing with Saussure on the arbitrariness of signs, claims that "grammar is symbolic in nature" (Langacker, 1987, p. 2), particularly through the role played by metaphors and metonymies (Ruiz de Mendoza & Otal Campo, 2002). The naming process based on [noun] + [noun] composition can be interpreted as a "recursive sub-classification of the head" (Martin, 1988, as cited in Ormrod, 2001, p. 10). In Langacker's (1987) words:

> "The schema describing the basic pattern for English compounds identifies the second member of the compound as the profile determinant: football thus designates a ball rather than a body part,

carrot juice names a liquid rather than a vegetable, blackbird is a noun rather than an adjective, and so on" (p. 290).

This means that, in a [noun] + [noun] construction, the head (i.e. the last noun) represents a whole domain, or category, and that the first noun defines the profile (a sub-category within that domain). An analogy can be drawn between this compounding process and metonymies, i.e. tropes in which a phrase is substituted for another closely related expression. Croft (2003) convincingly argued that a metonymy basically consisted in highlighting an active zone (a sub-category) within a domain and Ruiz de Mendoza and Diez (2003) have developed a clear graphical representation system by showing that all types of metonymies can be reduced to part-for-whole or whole-for-part relationships, representing this with an arrow pointing either to the whole or the part, see Figure 1. The examples they give include "She's taking the pill, where 'pill' stands for 'contraceptive pill'" and "All hands on deck, where by 'hands' we refer to sailors who do physical work in virtue of the hands playing an experientially prominent role" (Ruiz de Mendoza & Diez, 2003, pp. 496-497).

Figure 1. Diagrams inspired from Ruiz de Mendoza and Diez (2003, pp. 513-515)

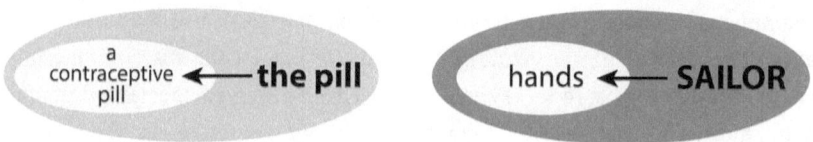

This figure could easily be adapted for CNs, for example football or carrot juice, see Figure 2.

If we admit there is an analogy between metonymies and [noun] + [noun] compounding, because both are based on a highlighting mechanism, and if the graphic design used for metonymies can be used for CNs as well, we now come to a third and last research question: would this type of illustration help students find the correct word order in a [noun] + [noun] construction?

Figure 2. Adaptation for CNs

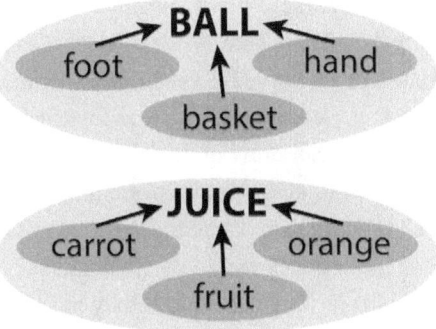

3. Methods

The case study reported below was carried out with process engineering students, because this Master's degree course covers a wide range of applied scientific topics. It has three main specialisms: energy engineering (including alternative sources of energy), environmental engineering (dealing with effluents and pollution), and chemical formulation (for the production of cosmetics or medicines). All the French students involved in this study were in the second year of their Master's degree. They had taken a computerised adaptive test[2] benchmarked on the CEFR on September 6th and were split into two comparable groups with the same English teacher. Sixteen students' results could be taken into account for this study (five learners at the A2 level and three at the B1 level in each group). The decision to take into account students' proficiency levels was motivated by a preliminary case study (Fries, 2015) in which the higher proficiency of students in the experimental group introduced a bias which made the findings difficult to interpret: were the results of the experimental group due to the way they were taught CNs, or to their better English level?

---

2. SELF (System of Evaluation for Languages with a Follow-up module for remediation). This adaptive test has been developed in Grenoble thanks to funds from the French national research agency.

Chapter 6

The control group was taught [noun] + [noun] compounds according to the Minimum Competence in Scientific English textbook: "nouns can also be modified by other nouns, i.e. these nouns function as if they were adjectives. [...] This explains why they do not take an 's' (with some exceptions), even after numerals" (Upjohn et al., 2013, pp. 138-139). In the experimental group, this was complemented with a cognitive-semantic approach focusing on the encyclopaedic meaning of words (students can draw on their specialised knowledge in order to understand CNs), the continuum between grammar and lexicon (CNs are also terms from specialised domains), and a graphic representation of [noun] + [noun] compounds as active zones within whole domains (see Figure 2 above, for example). This formal teaching was followed by practice exercises.

Then, during term time, both experimental and control groups followed a TBLT course which focused on note-taking and oral presentations. For both groups, the marked in-class assignments included written summaries of two videos dealing with energy transition in Denmark ('Smart energy systems: 100% renewable energy at national level'[3]) and with pharmaceutical formulation ('Fighting malaria with green chemistry: Artemisinin'[4]). For both groups, the final examination was an oral presentation on the internships they had done the previous year, in front of a jury comprising of a process engineering teacher and an English as a foreign language teacher. Although the teaching programme in both groups was exactly the same, the way the written assignments were corrected differed. In the control group, incorrect CNs were underlined and the letters 'CN' written in the margin (which meant there was a mistake to be corrected in the [noun] + [noun] construction), whereas students in the experimental group also had drawings with circles and arrows written in the margin to help them. Correct CNs were underlined in green in both groups.

The learner data used for this study are thus twofold. The first part includes the record of all the [noun] + [noun] constructions students used in the video-based summaries they wrote in class after the lesson they had on CNs, which

---

3. Podcast retrieved from https://www.youtube.com/watch?v=eiBiB4DaYOM
4. Podcast retrieved from https://www.youtube.com/watch?v=FfAJdnKqRCo

allows for a comparison of the performances of the two groups for the same tasks. The second part is a micro-corpus made up of the slideshows the students prepared for their final presentations, in which they were free to choose the words they wanted. In order to keep the two sets of data comparable, they have been restricted to written English, so the transcripts for the oral presentations themselves have not been included. The students were not asked to prepare a written version of their presentation, because it was felt counterproductive from a TBLT point of view: it would have encouraged them to read, instead of using public speaking skills.

4. Analysis and results

On the whole, both groups were able to grasp the CNs expressing the topic of each video and the key information given[5]. For the video on artemisinin, students in the control group as well as the experimental group identified the cause for the disease (malaria/plasmodium parasite), its target in the human body (blood cells, blood stream), and the name of the molecule which kills the parasite (hydrogen peroxide). However, students in the experimental group were much more precise in their use of CNs expressing a chemical reaction (extraction step, reaction step, solvent choices, water solvent...). In the second video, both groups were able to identify three key CNs: energy resources (wind turbines, wind energy/power), energy production (hydrogenation plants), and energy consumption (transport sector), but students in the control group were able to give a more detailed account of energy production (biogas plants, energy crops, fuel cells...). The tentative conclusions which can be drawn from these findings is that the textbook-based teaching given to all the students seems a sound basis to help them understand and reuse key CNs, with no clear added value coming from the cognitive perspective. In other words, it appears that the cognitive-semantic approach to teaching [noun] + [noun] constructions can at best complement, but in no way replace, a morphosyntactic perspective. In order to see more clearly what the benefit of the cognitive approach could be

---

5. See supplement, parts 1 and 2: https://research-publishing.box.com/s/80f8vaff3ohlyxdw8gd9pqh1db6t4y5q

in terms of implicit learning, appropriate style and word order, we now need to turn to the presentation slides. A concordance analysis of the slideshow corpus is shown in Table 1.

Table 1. Concordance analysis of the slideshows

	Experimental group (Cognitive-semantic approach)	Control group (Morphosyntactic approach)
Total number of words (tokens)	4,258	4,252
Word types	2,370	1,686
[noun] + [noun] compounds	214	152
CNs quoted more than 10 times in the Web of Science (WOS)	73	67
Within-group variation Min Max SD	2 19 4.8	3 16 4.4

The two-gram function of the Antconc concordancer[6] was used for the slideshow corpus and yielded 4,258 two-gram tokens and 2,370 two-gram types for the experimental group, versus 4,252 two-gram tokens and only 1,686 two-gram types in the control group, among which 214 were [noun] + [noun] constructions in the experimental group, versus 152 in the control group[7]. CNs were then checked in the WOS, a multi-disciplinary database of peer-reviewed journals, in order to compare the learners' writing with an expert corpus. All CNs quoted less than 10 times in the WOS were taken out. The results were 73 CNs for the experimental group and 67 in the control group, with wide variation across individuals. In the experimental group, one student produced only 2 CNs, while another produced 19, whereas in the control group, the range was 3 to 16, with standard deviations of 4.8 versus 4.4[8].

---

6. Concordancer retrieved from http://www.laurenceanthony.net/software/antconc/

7. Each of these constructions had to be checked manually in its context, because of formatting issues related to slide decks: Antconc does not take bullet points into account in its two-gram function. For example, an introductory slide containing four bullet points (Introduction - Results - Implementations - Conclusion), yielded "introduction results, results implementations and implementations conclusion".

8. See supplement, parts 2 and 3: https://research-publishing.box.com/s/80f8vaff3ohlyxdw8gd9pqh1db6t4y5q

Considering the small number of students involved in this case study and the extent of individual variation among them, the analysis that follows will focus on three qualitative elements: evidence for implicit learning of CNs, stylistic appropriateness for the required task, and reliability of the cognitive explanation given for word order in CNs. Implicit learning can be inferred from the presence of highly specialised CNs, referring to a technical knowledge coming directly from the student's internship, rather than from the English class. In the experimental group, these highly specialised CNs are drawn from the domains of chemistry (sodium carbonate, calcium aluminates), physics (phase diagram), spectroscopy (absorbance spectrum, transmittance spectrum), and fluid mechanics (vortex street). A definition is even provided by Student 4 in his slides: "A Karman vortex street is a repeating pattern of swirling vortices caused by the unsteady separation of a fluid around blunt bodies". In the control group, on the other hand, all the highly specialised CNs are chemical compounds (ammonia water, sodium valproate, tungsten carbide). This seems to indicate that students in the experimental group were able to translate the knowledge gained during their internship into [noun] + [noun] compounds more easily than in the control group.

Secondly, the special place of CNs in the middle of the continuum between grammar and lexicon might be able to account for their potential for compressing the meaning of a whole phrase into a single [noun] + [noun] construction. This is particularly useful for powerpoint slides, where whole sentences should be avoided as a rule and replaced by key points only. In the experimental group, almost half of the CNs were used on their own in a bullet point, whereas roughly three-quarters of the CNs written on the control group's slides were included in whole sentences. Constructions such as "failure modes and effect analysis", "weight assessment calculation", or "membrane separation technique" made the slides look as if the students really knew what they were talking about, while providing them with visual help for their oral presentations at the same time. The use of CNs in bullet points therefore worked in some measure as a compensation strategy, allowing students to hide their weaknesses in written English. Student 1, for example, who was creative in reusing CNs adequately and even coining elaborate new ones, such as "fruit juice phase diagram", also continued to make CN errors: "Works to comprehend the connection between food and its environment to develop generic tools for processes optimisation".

Finally, the large thematic scope of the CNs used by process engineering students on their slides allows us to check if the conceptualisation of a [noun] + [noun] compound as an active zone within a domain can really account for CNs in a wide range of scientific and technical fields. On the whole, mistakes in word order are infrequent in the slideshow corpus: only two in the experimental group ("area electrolysis" and "energy household") and three in the control group ("estimate energy losses", "storage ammonia", and "unit production"). All [noun] + [noun] compounds can be accounted for by the highlighting pattern, except a handful of figurative CNs, which are based either on a metaphor (cement plants, head office, key words) or on a metonymy (host organisation). The highlighting pattern therefore seems to be representative enough to be used as a teaching heuristic for CNs in S&T.

5. Conclusion

This exploratory case study of the use of CNs among French masters students in process engineering suggests that a cognitive-semantic approach to CNs could be beneficial to ESP teaching, as a complement to a morphosyntactic perspective. First of all, emphasis on the encyclopaedic meaning of words draws students' attention to the terminology of their specialised field of study. It therefore allows students to draw on their expertise in their specialised fields in order to understand CNs better. It also makes them aware of the fact that CNs are a clear example of the continuum between lexis and grammar, so that learning key lexicalised NPs will also help them to use [noun] + [noun] constructions correctly. This terminological perspective can also be used as a compensation strategy for slideshows, if students use more CNs and fewer verbs, thus adapting their scientific and technical writing to the appropriate register. Finally, a graphic representation of CNs as circles and arrows, though failing to capture a small number of figurative CNs, seems representative enough to guide students in choosing the right word order in the compounding process, so it seems useful, on the whole. These findings stand at the crossroads of didactics and ESP. Without drawing a distinction between explicit and implicit learning, it would be impossible to understand how students can use the terminology of their main fields of study correctly without studying it in their English classes. On the other hand, without taking into account specialised domains, it is quite difficult to understand why students should use CNs more accurately in fields of knowledge they are familiar with. However, this is only an exploratory case study, based on a very small learner corpus. It now needs to be followed by similar studies based on novice and expert corpora, for various fields.

6. Acknowledgements

Many thanks to Erin Cross, Jonathan Upjohn, and the two anonymous reviewers for their insightful comments.

References

Bauer, L. (1979). On the need for pragmatics in the study of nominal compounding. *Journal of Pragmatics, 3*(1), 45-50. https://doi.org/10.1016/0378-2166(79)90003-1

Bauer, L., & Renouf A. (2001). A corpus-based study of compounding in English. *Journal of English Linguistics, 29*(2), 101-123. https://doi.org/10.1177/00754240122005251

Benczes. R. (2011). Putting the notion of domain back into metonymy: evidence from compounds. In R. Benczes, A. Barcelona & F. J. Ruiz de Mendoza (Eds), *Defining metonymy in cognitive linguistics: towards a consensus view* (pp.197-216). John Benjamins. https://doi.org/10.1075/hcp.28.11ben

Berent I., & Pinker, S. (2007). The dislike of regular plurals in compounds: phonological familiarity or morphological constraint? *The Mental Lexicon, 2*(2), 129-181. https://doi.org/10.1075/ml.2.2.03ber

Biber D., & Gray B. (2016). *Grammatical complexity in academic English: linguistic change in writing.* Cambridge University Press. https://doi.org/10.1017/CBO9780511920776

Cabre, M. T. (1999). *Terminology, theory, methods and applications.* John Benjamins. https://doi.org/10.1075/tlrp.1

Council of Europe. (2001). *Common European framework of reference for languages*. Cambridge University Press.

Croft, W. (2003, reprinted). The role of domains in the interpretation of metaphors and metonymies. In R. Dirven & R. Pörings (Eds), *Metaphor and metonymy in comparison and contrast* (pp. 161-205). Mouton de Gruyter.

Downing, P. (1977). On the creation and use of English compound nouns. *Language, 53*(4), 810-842. https://doi.org/10.2307/412913

Ellis, R. (2003). *Task-based language learning and teaching*. Oxford University press.

Ellis, R. (2005). Measuring implicit and explicit knowledge of a second language: a psychometric study. *Studies in Second Language Acquisition, 27*(2), 141-172. https://doi.org/10.1017/S0272263105050096

Fries, M. H. (2015, May). *Approche cognitive de l'enseignement des noms composés en anglais : l'exemple des étudiants en génie des procédés*. Paper presented at the AFLICO conference, Grenoble, France.

Fudge, E. (1984). *English word-stress*. Allen and Unwin.

Giegerich, H. J. (2004). Compound or phrase? English noun-plus-noun constructions and the stress criterion. *English Language and Linguistics, 8*(1), 1-24. https://doi.org/10.1017/S1360674304001224

Hulstijn, J. H. (2005). Theoretical and empirical issues in the study of implicit and explicit second-language learning. *Studies in Second Language Acquisition, 27*(2), 129-140. https://doi.org/10.1017/S0272263105050084

Krashen, S. D. (1981). *Second language acquisition and second language learning*. Pergamon.

Krashen, S. D. (1987). *Principles and practices in second language acquisition*. Prentice-Hall.

Langacker, R. (1987). *Foundations of cognitive grammar* (Vol 1). Stanford University Press.

Lees, R. B. (1963). *The grammar of English nominalisations*. Indiana University Press.

Levi, J. N. (1978). *The syntax and semantics of complex nominals*. Academic Press.

Maniez, F. (2007). Using the web and computer corpora as language resources for the translation of complex noun phrases in medical research articles. *Panace@, 9*(26), 162-167.

Martin, J. R. (1988). Hypotactic recursive systems in English: towards a functional representation. In J. D. Benson & W. S. Greaves (Eds), *Systemic functional approaches to discourse* (pp. 240-270). Ablex Publishing Corporation.

Olsen, S. (2000). Compounding and stress in English: a closer look at the boundary between morphology and syntax. *Linguistische Berichte, 181*, 55-70.

Ormrod, J. (2001). Construction discursive de noms composés dans des textes scientifiques anglais. In D. Banks (Ed.), *Le groupe nominal dans le texte spécialisé* (pp. 9-23). L'Harmattan.

Parkinson, J. (2015). Noun–noun collocations in learner writing. *Journal of English for Academic Purposes, 20*, 103-113. https://doi.org/10.1016/j.jeap.2015.08.003

Rossi, C., Frérot, C., & Falaise A. (2016). Integrating controlled corpus data in the classroom: a case-study of English NPs for French students in specialised translation. In F. A. Almeida, L. Cruz Garcia & V. Gonzalez Ruiz (Eds), Corpus-based studies on language varieties (pp. 167-190). Peter Lang.

Ruiz de Mendoza, J., & Diez, I. (2003, reprinted). Patterns of conceptual interaction. In R. Dirven & R. Pörings (Eds), *Metaphor and metonymy in comparison and contrast* (pp. 489-532). Mouton de Gruyter.

Ruiz de Mendoza, J., & Otal Campo, J. (2002). *Metonymy, grammar and communication.* Comares.

Ryder, M. E. (1994). *Ordered chaos: the interpretation of English noun-noun compounds.* University of California Press.

Salager-Meyer, F. (1984). Compound nominal phrases in scientific-technical literature: proportion and rationale. In A. K. Pugh & J. M. Ulijin (Eds), *Reading for professional purposes – Studies and practices in native and foreign languages* (pp. 138-149). Heinemann.

Torres, F. A. (2015). A classroom-centred approach to the translation into Spanish of common noun compounds phrase patterns in English technical texts. *Revista de Lenguas para Fines Específicos, 4*, 13-24.

Upjohn J., Blattes S., & Jans, V. (2013, reprinted). Minimum competence in scientific English. Editions de physique.

Wilches Alvear, R. E. (2016). *Compound nouns and reading comprehension: creating awareness of compound nouns to improve the reading comprehension of students of English at the university of Cuenca.* Master thesis. University of Cuenca, Ecuador.

7 When storytelling meets active learning: an academic reading experiment with French MA students

Pauline Beaupoil-Hourdel[1], Hélène Josse[2], Loulou Kosmala[3], Katy Masuga[4], and Aliyah Morgenstern[5]

Abstract

Reading, understanding, analysing, and synthesising texts in English are skills all French university students in humanities and social sciences are expected to develop. "Research and Storytelling" is a pedagogical project funded by Sorbonne Paris Cité that aims to help Master of Arts (MA) students in the humanities become better readers of research articles by using a specific set of narrative devices. We hypothesised that the use of storytelling devices would not only improve comprehension of scientific articles but also ease anxiety and raise confidence. The data provided by 26 students is composed of 11 journal entries per student (one per class session), a stress test completed at the end of the semester, their grades and feedback, and their final evaluation of the course. Results indicated that after 12 two-hour sessions, most students had a sharper and more robust method for reading articles than before the course. They felt more comfortable with the task, and their self-confidence had increased. Most students

---

1. Sorbonne Nouvelle University, Paris, France; pauline.beaupoil-hourdel@espe-paris.fr

2. Sorbonne Nouvelle University, Paris, France; helene.josse@sorbonne-nouvelle.fr

3. Sorbonne Nouvelle University, Paris, France; loulou.kosmala@gmail.com

4. Sorbonne Nouvelle University, Paris, France; masuga@uw.edu

5. Sorbonne Nouvelle University, Paris, France; aliyah.morgenstern@sorbonne-nouvelle.fr

How to cite this chapter: Beaupoil-Hourdel, P., Josse, H., Kosmala, L., Masuga, K., & Morgenstern, A. (2017). When storytelling meets active learning: an academic reading experiment with French MA students. In C. Sarré & S. Whyte (Eds), *New developments in ESP teaching and learning research* (pp. 109-129). Research-publishing.net. https://doi.org/10.14705/rpnet.2017.cssw2017.748

Chapter 7

also stated that after taking the course they felt more able to re-inject the knowledge and know-how they had acquired into the writing of their master's thesis.

Keywords: storytelling, narratives, scientific reading, motivation, pragmatic skills.

1. Introduction

French university lecturers often complain that their graduate students are poorly equipped for reading research papers in English. As advocated by Turner (1996), reading scientific writing as narrative might help students become better and more frequent readers of research papers. Storytelling can be used as a natural scaffolding provided by human cognition to bring structure to challenging information; it activates the whole brain as it involves mental simulation and imagery.

In the framework of a pedagogical project, we devised a course to help MA students become better readers of research articles by using a specific set of narrative devices and an interactive pedagogical approach. In this chapter, we review previous studies, present our devices, and the results of an experiment designed to test their impact on academic reading by students in the project.

2. Students' reading habits and academic demands

Reading, understanding, and synthesising texts in English are skills almost all French university students are expected to develop. Yet, at the undergraduate level, students are rarely taught how to read texts in English as a Foreign Language (EFL) (Hill, Soppelsa, & West, 1982; Spack, 1988). Additionally, skills developed during undergraduate years in English grammar, vocabulary, and syntax do not easily transfer to scientific articles (Luna, 2013; Schuls, 1981). Students are thus expected to learn how to read complex scientific content on their own and to be able to synthesise and analyse it, based on their ability to read English. Yet the ability to read or speak English does not automatically entail that one can read and understand scientific articles written in English. Students need to be taught those skills (Ro, 2016).

Reading academic papers involves both reading skills and the mastery of a large vocabulary (Laufer & Sim, 1985). However, over the last few decades, scholarly research and governmental and media surveys have described what has been called a "crisis" in young people's reading habits. In 2007, the NEA survey has reported a constant decline in numbers of daily readers and a significant increase in non-readers since the 1980's. In France, similar surveys drew the same conclusions (Donnat, 1998, 2008). Baudelot, Cartier, and Détres (1999) showed that young adults read the books required for school, but most teenagers stop reading for pleasure after the age of fifteen and begin to consider reading boring. One explanatory factor could be that the books they are assigned at school do not interest them. Another factor is that reading is not considered a socialising activity and, in the era of the internet and social networking, young adults turn more easily to electronic devices and social interaction online (Donnat, 2012).

Previous research on the attitude of university students towards reading has shown difficulty in drawing an objective picture: students now read more online than they read with printed material (Collège scientifique de l'OVE, 2010), but Lahire (2002) observed that fewer than 50% of students read assigned material from beginning to end and concluded that students do not read assigned material to expand their knowledge on a topic but rather to find specific information.

According to Lacôte-Gabrysiak (2015), humanities students enjoy fantasy books because it reminds them of the stories they used to enjoy when they were young. Their taste for non-academic reading is therefore guided by their attraction to narratives. And yet, Krashen (2004) noted that the opportunity for pleasure reading is often missing in EFL classroom contexts. Storytelling can thus be used to introduce a dimension of pleasure into academic reading. Previous research has shown that narratives are easier to comprehend, and audiences find them more engaging than traditional "logical" scientific communication (Graesser,

Olde, & Klettke, 2002; Green, 2006). Narratives are associated with increased memory and better understanding (Moore, 1999; Schank & Abelson, 1995), and several researchers have advocated the use of narratives as a learning format (Dahlstrom, 2012; Reiss, Millar, & Osborne, 1999). For all those reasons, it appears that storytelling can be a powerful educational tool useful for training graduate students to read complex scientific papers and develop a critical mind more easily.

3. The "Research & Storytelling" project

3.1. Description of the project

The research cited above shows that students are often prevented from engaging in scientific reading because of a lack of daily practice and taste for this solitary occupation and particular type of literature, along with low levels of motivation. In a survey we conducted on our students in the humanities, five out of 26 answered they did not like reading either books or academic papers. The *Science & Storytelling Project*[6] was launched in September 2015 to design a course on academic reading that would tackle the reading obstacles described in the literature. Our contention was that an academic reading course would be most efficient when providing not only reading devices that would rely on students' individual tastes for storytelling but also an active learning environment that would appeal to their socialising skills and fuel their motivation. Our aim was not merely to improve the students' reading skills, but to develop their self-efficacy beliefs (Bandura, 1997; Pajares, 1996) with the longer-term goal of becoming efficient and regular academic readers.

There exist various types and forms of scientific articles. Because we opted for a hands-on course, we decided not to linger on this variety. To match our students' needs, we focussed on scientific articles in the areas of the arts and humanities.

---

6. The project, launched by Pr. Aliyah Morgenstern influenced by an original idea by Monica Gonsales-Marques, benefited from an Idex Sorbonne Paris Cité funding.

The main objective of the project is to design and test an academic reading course based on storytelling that could be taught to different audiences (MA students, Ph.D. students) in different formats (24-hour, six-hour and three-day courses).

With this goal in mind, the following research questions guide the present chapter:

- How do storytelling and the reading devices affect our EFL students' academic reading skills?

- How do storytelling and the reading devices affect our EFL students' academic reading motivation?

- Does the programme set the students on the path of reflexive thinking?

3.2. The MA reading course under study

The course was taught over two years to 148 students, at MA (n=76) and Ph.D. (n=72) levels. This study concentrates on the 24-hour course designed for MA students majoring in English Studies and taught in 2016-17, during the second year of the experiment at Sorbonne Nouvelle University. Although the given course was taught to English majors, the project meets the characteristics of the research framework developed for English for specific purposes by Sarré and Whyte (2016) as it relies on the interaction between language, content knowledge, and methodology of a specific domain (Douglas, 2010). The objectives of the course were announced to the students during the first session (Figure 1).

The students were told that active participation was expected, as interaction with discussion sessions and mutual learning in groups works better towards the achievement of the objectives. To allow students greater autonomy, the instructor also planned role plays: students were asked to review abstracts and articles for a peer-reviewed journal; they impersonated in turn the role of a research supervisor as they listened to their peers present their research projects and gave their fellow students constructive feedback. Teacher-fronted slide shows were kept to a minimum and an inductive method was preferred to introduce the various storytelling reading devices: these were presented after the students had brainstormed and discussed their findings in groups.

Figure 1. Objectives of the course

The course was planned and the material designed to encourage and sustain a reflective approach (Gokhale, 1995; Poteaux & Berthiaume, 2013). Students were required to complete a weekly journal for each of the 11 sessions. The format and questions were the same for each session:

- summarise the session in three short sentences;

- give a title to the session;

- indicate how this session changed you in terms of knowledge, know-how and social/personal skills;

- what did you prefer in this class? Briefly explain your answer.

The instructor wrote a journal entry at the end of each session to reflect upon the teaching material and the teaching experience.

Overall, 26 students attended the course, 22 women and four men aged 21-28. Among them 21 declared that their mother tongue was French, one English, one Arabic, one Russian and two Italian. Ten students were first year MA students and had never read scientific articles, while 16 were second year MA students who had already written a first-year thesis and had read scientific material. The students were all English majors, and their level of English ranged from B2 to C2[7].

3.3. The data available

The data is composed of the class grades assigned on the basis of an assessment grid, their journal entries, the instructor's teaching journal entries, scores on a stress test completed at the end of the semester, the graded papers, and a feedback survey of 27 questions answered online at the end of the semester (Table 1). We collected 50 graded papers (home assignment and final exam) as one of the students was a guest student.

Table 1. Overview of the data

Nature of the data	Number
Stress test (week 12)	26
Home assignment (graded)	25
Final exam	25
Students' journals	234
Teacher's journal	11
Feedback survey	26

---

[7]. We evaluated the students' level following the common European framework of reference for languages, Council of Europe (2001).

4. Storytelling devices to inform academic reading

The devices presented in this section already exist in the literature. However, they were remodelled and adapted as most of them were designed as writing devices and were also not specifically aimed at EFL students. We presented five devices so that the students could choose which one best fitted their needs and the scientific article under study. Having the students select and defend their preferred device was part of the empowering process we tried to sustain.

Narrative Elements: This device is derived from Luna (2013). The author lists the narrative elements present in any tale and claims that they can be equally found in research articles. The Narrative Elements include *Protagonist, Antagonist, Scene, Conflict, Stakes,* and *Resolution*. Luna's contention is that being aware of this similarity makes students better scientific writers. We tried to show our students that it makes them more efficient readers because looking for these elements within a scientific article provides them with a map to find their way through the article and fuels their agentivity. Figure 2 and Figure 3 show how both a tale and the title of a scientific article (Bastian, Jetten, & Fasoli, 2011) can be presented.

Figure 2. The narrative elements in *Little Red Riding Hood*

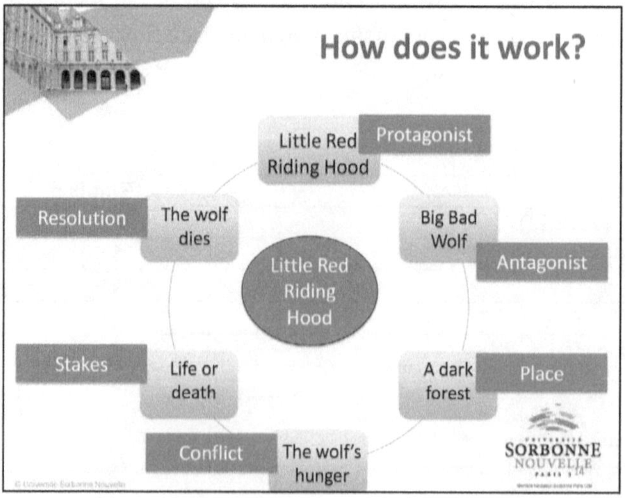

Figure 3. The narrative elements in the title of a scientific article

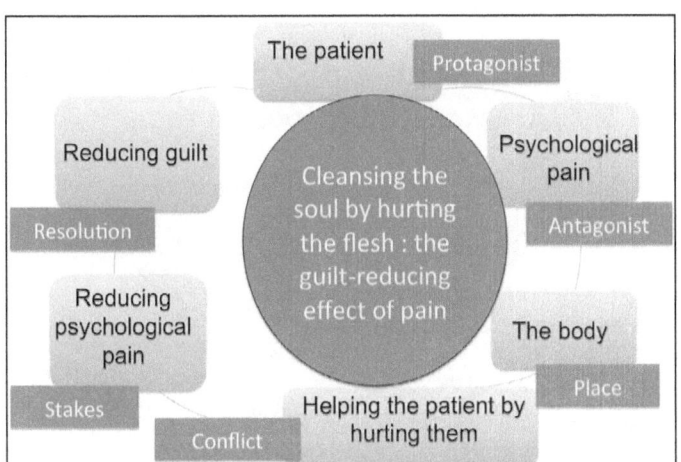

Narrative Spectrum – And... But... Therefore (ABT) structure: This device is derived from Olson (2015). He points out that a story – or a scientific article – is not an accumulation of unrelated facts. To be compelling, both stories and scientific articles should present facts along a Hegelian dialectic consisting of 'and', 'but', and 'therefore'. Thus, we tried to teach our students to identify this structure in scientific articles for them to understand the underlying logic. The device was first tested using a tale: "Little Red Riding Hood is asked to bring her grandmother some food AND not to talk to strangers BUT she tells the wolf more than he needs to know, triggering off her untimely death, THEREFORE little girls should listen to their mother's advice".

Dramatic Arc: This device is derived from Luna (2013). With the "Dramatic Arc", the author places emphasis on the way tension builds and subsides in storytelling. This writing device is about the momentum a scientific article should gather so as to help the reader follow the argument more easily. As a reading device, the "Dramatic Arc" provides the students with yet another structure to follow the author's thinking and to discriminate among the arguments and the results. In Figure 4, we applied the device to *Little Red Riding Hood*.

Figure 4. The dramatic arc, adapted from Luna (2013)

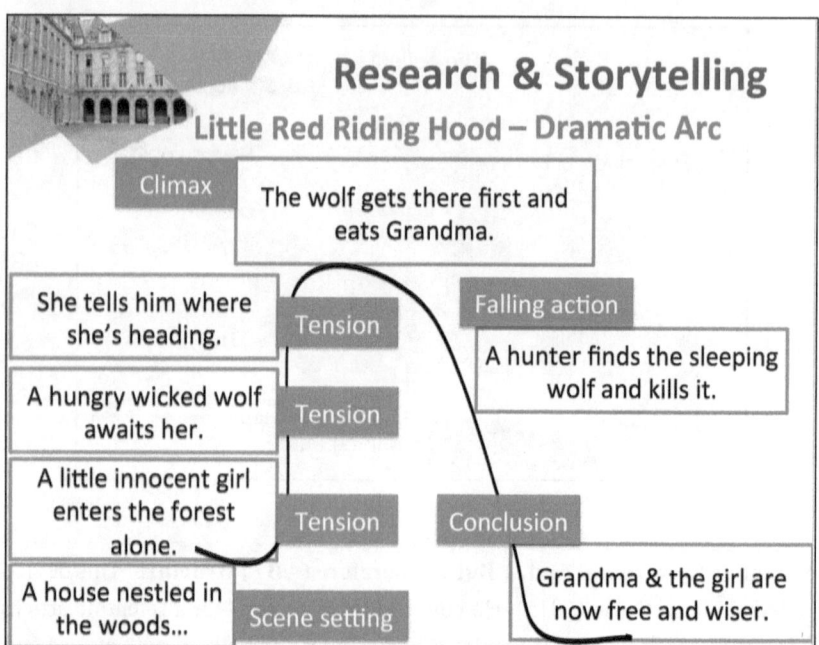

Introduction, Method, Results, Analysis, and Discussion (IMRAD): This refers to the prototypical structure of a scientific article (based on experimental psychology papers) that is commonly taught in academic writing courses (Olson, 2015). We chose to present the device although it is rarely used as a reading device in the humanities.

Recall Diagram: This device is derived from Smith and Morris (2014). It differs from the other four devices presented above as it is useful for synthesising and retaining information and not necessarily for guiding the students through the articles. The theory behind the 'recall diagram' is that readers retain information better if they set it in relation to their own lives. The recall diagram (Figure 5) is a reading, not a writing device. It is not a storytelling device in itself but it implies that scientific articles tell stories, and as such, we may relate to them just as we relate to stories.

Figure 5. Recall diagram adapted from Smith and Morris (2014)

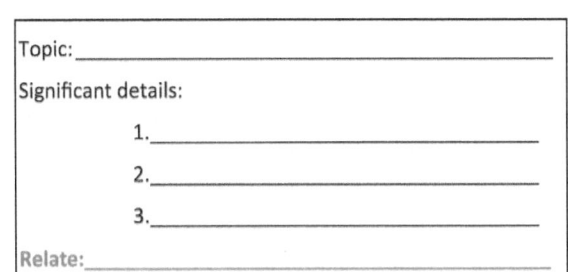

The instructor drew the students' attention to the distinction between the 'relate' and 'react' sections. The 'react' section is the reader's critical response to the article, whereas the 'relate' section is setting the article in relation to his/her personal life. This section is crucial due to the hypothesis that readers better retain an article's main ideas when personally affected by its contents, both positively or negatively. The students were thus invited to always keep track of their feelings when reading scientific materials, similarly to the way a young child can experience fear, sadness, or happiness when reading tales.

5. Results and discussion

In this section we address the research questions and assess whether the students developed reading skills and reflexive thinking. We also assessed whether the course fueled their motivation.

5.1. Learning outcomes

5.1.1. Getting the students to read more

One of the goals of the course was to get students to read academic work. Overall, during the semester, the students read eight articles, ten abstracts, and one fictional narrative, for a total number of 63,339 words, i.e. about 5,280 words per week. The general survey completed by the students online at the end of the semester offers a favorable insight into the students' involvement in the course, as only one student admitted he had rarely done his homework, whereas 16 answered that they often did their homework, and nine claimed that they always did their homework. The students never discussed workload in their journal entries, and the instructor never wrote that she had the feeling that not all the students did their reading assignments. Rather, she explained that the students were eager to answer her questions and to participate in class.

5.1.2. Getting the students to read better

The two graded papers were designed to evaluate whether the main objectives of the course were achieved. The papers tested the students' ability to use the reading devices, synthesise articles, and keep a concise record. Table 2 illustrates their proficiency for all these skills.

Table 2. Students' proficiency in reading, synthesising and keeping record of scientific articles at the end of the semester (n=25)

	Using the reading tools	Synthesising	Keeping a record
Acquired	12	9	3
In process	10	13	17
Not acquired	3	3	5

Results show that half the students (12) had acquired the use of the reading devices after a semester of training, and close to the other half (10) were in the process of acquiring it.

Some of the students seemed aware of the empowering dimension of the reading devices. In a journal entry, Kevin[8] explicitly drew a parallel between storytelling

---

8. We changed the names of the students to guarantee anonymity and retained only the gender. Therefore Kevin refers to a man.

and the structure of scientific articles. He explained that focussing on the structure of an article helped him access its contents and retain it:

> "The storytelling elements can be a very helpful tool to analyse the parts of an academic text, helping the reader to identify the different parts of an article and assimilate its content in a more efficient manner. For me, it has made the text less abstract and it has changed my way to deal with the text"[9] (Kevin).

Interestingly, the students who know how to use the devices are more likely to synthesise the content of articles accurately: out of the 25 students, 22 either acquired or were in the process of acquiring the synthesising skill. However, only nine students were fully able to synthesise articles at the end of the semester, while more than half of the class still had to improve. Students did not keep a record of their scientific reading before the course. The skill was thus completely new to them. However, if only three students had full mastery of the recording skill, 17 more were in the process of acquiring it. We also observed that the three students who could write a record by the end of the course were among the nine who could synthesise and the 12 who used the reading devices correctly. Keeping a record is among the latest stages in Bloom's (1956) taxonomy and before creating a report, students need to read articles in an efficient way, identify key notions, synthesise them, and organise information. Writing a record, therefore, is achieved by building upon other skills.

5.1.3. Setting the students on the path of critical reading

Throughout the semester, the students were trained to become more reflective readers. In her journal entry, Melissa wrote about the reviewing activity:

> "I liked the reviewing part because it forces us to take some distances from the paper […]. I think it makes us realise that producing a scientific

---

[9]. The students' comments were not corrected for errors in English.

paper, in the structuring part essentially, is all about choices, which can be questioned".

The students were encouraged to reflect upon their learning experience and their mastery of the narrative devices in their journals. We double blindly-coded the the students' comments (n=234) and counted the number of journals in which the students wrote reflective comments on the course and/or their learning experience. The agreement between the two raters was high (97%, 226/234, Cohen's Kappa, κ = 0.94). We classified the students into three categories:

- not reflective: the students adopting a reflective stance in fewer than two journal entries;

- somewhat reflective: the students adopting a reflective stance in four to six journal entries;

- reflective students : the students adopting a reflective stance in seven or more journal entries (see Table 3).

Table 3.   Distribution of the reflective students throughout the semester

Not reflective students	Somewhat reflective students	Reflective students
5	13	7

When the course began, students were not used to engaging in critical thinking. It was only after four weeks of training that reflective remarks started to emerge in the journals. The data also indicates that the seven students who wrote critical remarks in their journals were always the same students, M1 and M2 alike. Overall, 20 students wrote reflective comments in their journals. These results indicate that thinking critically is a difficult skill to acquire.

5.2. Shift in attitude

Since becoming a regular reader of articles in a foreign language cannot be achieved overnight, the course was designed mostly to change the students' attitude towards academic reading in English so that they might in the future evolve into regular critical academic readers. An important part of this study was thus to monitor any shift in attitude.

5.2.1. Motivational factors

In the first three journal entries, few students wrote about motivation, yet after Session 4, eight to 13 students per session spontaneously wrote about their motivation to engage with the course (Table 4).

Table 4. Number of students who identified factors of increasing (+) or decreasing (-) motivation in their journal entries

	Motivation +	Motivation -
Journal 1	6	1
Journal 2	3	1
Journal 3	4	0
Journal 4	10	0
Journal 5	10	0
Journal 6	6	0
Journal 7	13	0
Journal 8	9	0
Journal 9	9	0
Journal 10	5	0
Journal 11	11	0

This development of motivation can be observed in Marion's comments. In Journal 2, she wrote that she preferred working on her own and did not like group work: "I prefer to work alone and then explain my thoughts to the other students instead of developing an answer in group" (Marion).

But in Journal 4, she changed her attitude towards the teaching method that constantly relied on interaction to collectively build knowledge:

"I think that this way of learning, that gives a lot of importance to practical approach and to the sharing of our thoughts, is really helpful to me, because I learn very fast and I can immediately apply the new-learnt skills in my reading activity" (Marion).

The students were not directly encouraged to mention motivation in their journals. The fact that eight to 13 chose to do so after Session 4 is noteworthy information on the learning experience of the students in this course.

In the end-of-semester survey of the course, the students were asked to express whether they had identified any factors of motivation that encouraged their participation in class. Fifteen students responded in the affirmative, selecting "interaction" as the most important factor, followed by their desire to improve and learn. It appears that our teaching approach, along with the use of storytelling to read articles, enhanced motivation and perceived knowledge acquisition.

5.2.2. Stress factors

To get a better overview of the results on motivation, we looked for occurrences of strings related to stress in the journal entries, as shown in Table 5. We observe a decrease of the stress (stress - category) after Session 11, hence a week before the final exam. In Journal 11, 11 students wrote about stress; one mentioned that she was afraid to take the exam (stress +) whereas ten spontaneously wrote that they felt prepared for the task and that they were confident (stress -). The other 14 students did not mention stress.

Table 5. Number of students who identified factors increasing (+) or decreasing (-) stress in their journals

	Stress +	Stress -
Journal 1	2	3
Journal 2	4	2
Journal 3	0	1
Journal 4	0	2
Journal 5	1	2
Journal 6	0	0

Journal 7	0	3
Journal 8	0	0
Journal 9	1	2
Journal 10	3	2
Journal 11	1	10

Just before the final exam, the students filled out a stress test (Table 6) and 22 students answered they felt confident before starting the final exam. All the students felt that they had the knowledge and skills to do the exam and all but one felt prepared for the task.

Table 6. Results to the stress test answered before the final exam

	Not at all	A little	Somewhat	Moderately	Very	Extremely
1. I am confident that I know how to approach this task	0	1	3	14	8	0
2. I feel I have the knowledge and skills to do this task well	0	0	0	16	9	1
3. I feel well prepared for this task	0	0	1	14	11	0

These results corroborate the tendencies we observed in the journal entries; by the end of the semester, the students felt that they were skillful enough to read scientific articles for their master's' degree.

6. Conclusion

In this chapter we presented a new course for MA students we developed as part of the *Science and Storytelling Project*. The main objective was to train graduate students to become good readers of scientific papers. We conveyed qualitative analyses of the 2016 Master cohort (n=26) in the English Department at Sorbonne Nouvelle University, relying on 234 students' journal entries, 11 teacher's journals, a stress test, the students' exams, and a feedback survey.

Chapter 7

Our hypothesis was that providing a course using storytelling and reading devices to teach academic reading would help our students understand academic papers while giving them pleasure to read complex material in English.

We assessed the impact of the course following our research questions. We analysed whether storytelling and reading devices affected our EFL students' academic reading skills and motivation, and helped them develop reflective thinking. Results showed that the course managed to reduce anxiety and to raise motivation. By using storytelling as the basis of the learning process, by sustaining it with active teaching methods and by providing them with sizable material to read every week, the programme better prepared students to face academic reading requirements in English.

Yet, if preliminary results tend to indicate that half of the students started to think reflectively at the end of the semester, we were not able to draw significant conclusions regarding the acquisition of academic reading skills. Indeed, by the end of the semester most students were still acquiring the methods and skills needed to synthesise or keep a record of a scientific paper.

Although the main goal of the course has been achieved (i.e. putting the students on the path to critical thinking and reading), it also appears that this study should be monitored over a longer period. This pilot study cannot sufficiently corroborate the students' progress and improvements. Indeed, the course lowered the students' stress and boosted their motivation, but we were not able to measure whether it helped them develop their reading skills and become more autonomous readers.

For future research, it would be interesting to investigate which devices were preferred and why, looking at several independent variables such as the profile of the learner, their field of study, their proficiency, and their taste for reading. Moreover, improved reading comprehension can benefit other areas of learning and specially writing. Since writing involves imitation, implementing a programme such as the *Research and Storytelling* method might have the potential to improve academic writing. Overall, the programme has given the students the devices they needed to understand the subject matter, provided them with intensive and efficient practice, and has proven to make students read more eagerly.

7. Acknowledgements

We would like to thank the two editors of this volume, Shona Whyte and Cédric Sarré, as well as the two anonymous reviewers for their insightful and constructive suggestions. We also thank Tamar Kremer-Sadlik for her helpful, invaluable advice and for our rich discussions throughout the *Science and Storytelling Project*.

References

Bandura, A. (1997). *Self-efficacy: the exercise of control* (1st edition). Worth Publishers.

Bastian, B., Jetten, J., & Fasoli, F. (2011). Cleansing the soul by hurting the flesh: the guilt-reducing effect of pain. *Psychological Science, 22*(3), 334-335. https://doi.org/10.1177/0956797610397058

Baudelot, C., Cartier, M., & Détres, C. (1999). *Et pourtant ils lisent...* Seuil.

Bloom, B. S. (1956). *Taxonomy of educational objectives book 1: cognitive domain* (2nd edition edition). Addison Wesley Publishing Company.

Collège scientifique de l'OVE. (2010). *Observatoire national de la vie etudiante* (Enquête « Conditions de vie des étudiants en France » (CdV)).

Council of Europe. (2001). *Common European framework of reference for languages: learning, teaching, assessment*. Cambridge University Press.

Dahlstrom, M. F. (2012). The persuasive influence of narrative causality: psychological mechanism, strength in overcoming resistance, and persistence over time. *Media Psychology, 15*(3), 303-326. https://doi.org/10.1080/15213269.2012.702604

Donnat, O. (1998). *Les pratiques culturelles des Français, enquête 1997*. La Documentation Française.

Donnat, O. (2008). *Les pratiques culturelles des Français à l'ère numérique, enquête 2008*. Ministère de la culture et de la communication.

Donnat, O. (2012). La lecture régulière de livres: un recul ancien et général. *Le Débat, 3*, 42-51. https://doi.org/10.3917/deba.170.0042

Douglas, D. (2010). This won't hurt a bit: Assessing English for nursing. *Taiwan International ESP Journal, 2*(2), 1-16.

Gokhale, A. A. (1995). Collaborative learning enhances critical thinking. *Journal of Technology Education, 7*(1). https://doi.org/10.21061/jte.v7i1.a.2

Graesser, A. C., Olde, B., & Klettke, B. (2002). How does the mind construct and represent stories. In M. C. Green, J. J. Strange & T. C. Brock (Eds), *Narrative impact: social and cognitive foundations* (pp. 229-262). L. Erlbaum Associates.

Green, M. C. (2006). Narratives and cancer communication. *Journal of communication, 56*(s1), S163-S183. https://doi.org/10.1111/j.1460-2466.2006.00288.x

Hill, S. S., Soppelsa, B. F., & West, G. K. (1982). Teaching ESL students to read and write experimental-research papers. *TESOL Quarterly, 16*(3), 333-347. https://doi.org/10.2307/3586633

Krashen, S. D. (2004). *The power of reading: insights from the research* (2nd edition). Heinemann.

Lacôte-Gabrysiak, L. (2015). La lecture des étudiants entre plaisir et contraintes. *Communication. Information médias théories pratiques, 33*(1).

Lahire, B. (2002). Formes de la lecture étudiante et catégories scolaires de l'entendement lectoral. *Sociétés contemporaines, 4*(48), 87-107. https://doi.org/10.3917/soco.048.0087

Laufer, B., & Sim, D. D. (1985). Measuring and explaining the reading threshold needed for English for academic purposes texts. *Foreign Language Annals, 18*(5), 405-411. https://doi.org/10.1111/j.1944-9720.1985.tb00973.x

Luna, R. E. (2013). *The art of scientific storytelling*. Amado International.

Moore, D. (1999). Influence of text genre on adults' monitoring of understanding and recall. *Educational Gerontology, 25*(8), 691–710. https://doi.org/10.1080/036012799267440

NEA. (2007). *National Endowment for the Arts, Annual Report*. NEA.

Olson, R. (2015). *Houston, we have a narrative: why science needs story*. University of Chicago Press. https://doi.org/10.7208/chicago/9780226270982.001.0001

Pajares, F. (1996). Self-efficacy beliefs in academic settings. *Review of Educational Research, 66*(4), 543-578. https://doi.org/10.3102/00346543066004543

Poteaux, N., & Berthiaume, D. (2013). Comment soutenir les apprentissages des étudiants. In D. Berthiaume & N. Rege (Eds), *La pédagogie de l'enseignement supérieur : repères théoriques et application pratiques. Tome 1 : enseigner au supérieur* (pp. 39-54). Berne.

Reiss, M. J., Millar, R., & Osborne, J. (1999). Beyond 2000: science/biology education for the future. *Journal of biological education, 33*(2), 68–70. https://doi.org/10.1080/00219266.1999.9655644

Ro, E. (2016). Exploring teachers' practices and students' perceptions of the extensive reading approach in EAP reading classes. *Journal of English for Academic Purposes, 22,* 32-41. https://doi.org/10.1016/j.jeap.2016.01.006

Sarré, C., & Whyte, S. (2016). Research in ESP teaching and learning in French higher education: developing the construct of ESP didactics. *ASp (69),* 139–1164.

Schank, R., & Abelson, R. P. (1995). Knowledge and memory: the real story. *Advances in social cognition, 8,* 1-85.

Schuls, R. A. (1981). Literature and readability: bridging the gap in foreign language reading. *The Modern Language Journal, 65*(1), 43-53. https://doi.org/10.1111/j.1540-4781.1981.tb00952.x

Smith, B. D., & Morris, L. (2014). *Bridging the gap: college reading* (11th Edition). Pearson.

Spack, R. (1988). Initiating ESL students into the academic discourse community: how far should we go? *TESOL quarterly, 22*(1), 29-51.

Turner, M. (1996). *The literary mind: the origins of thought and language.* Oxford University Press.

Section 3.

Moving ahead: towards new practices

8 The SocWoC corpus: compiling and exploiting ESP material for undergraduate social workers

Jane Helen Johnson[1]

Abstract

Successful social work is based on communication (Clark, 2000; Pierson & Thomas, 2000; Thompson, 2010), yet there is little consensus in the field as to what exactly constitutes proper communication (Richards, Ruch, & Trevithick, 2005; Trevithick et al., 2004) and where models of good communication might be found. It is therefore not surprising that there is little material available for teaching the language required for social services students studying English for Specific Purposes (ESP) at undergraduate level in European universities (Kornbeck, 2003, 2008). This chapter describes the creation of a specialised corpus of material related to social work to meet the needs of undergraduate students using L2 English in Italy. It focuses on specific needs of this population (such as the language of social work and recent migration patterns), then describes the selection and preparation of source material for a monitor corpus in this domain. The chapter then shows how the resulting corpus can be used by teachers to select samples and analyse texts for classroom discussion. In conclusion, it is suggested that such material will contribute to language development through awareness-raising in the classroom as regards the effects of different language choices in a discourse context, while also mentioning areas for further development.

Keywords: SocWoC corpus, undergraduate social workers, ESP material.

1. University of Bologna, Bologna, Italy; janehelen.johnson@unibo.it

How to cite this chapter: Johnson, J. H. (2017). The SocWoC corpus: compiling and exploiting ESP material for undergraduate social workers. In C. Sarré & S. Whyte (Eds), *New developments in ESP teaching and learning research* (pp. 133-151). Research-publishing.net. https://doi.org/10.14705/rpnet.2017.cssw2017.749

1. Introduction

Growing specialisation and the need for awareness of how language works within particular domains has led to an increasing demand in Europe for courses in ESP at tertiary level. In the social services field, the lack of material for teaching ESP to social services students at undergraduate level has long been recognised (Kornbeck, 2003, 2008). Since it is widely acknowledged that appropriate language and discourse are fundamental for successful social services work (Thompson, 2010), this is no small problem. In order to fill this gap, it is essential to have a clear view of what is meant by 'appropriate language and discourse' in this field and be able to relate this to the type of language competence students need to do social work in their local context, since "[s]ystematic knowledge about language and practical awareness of how it works is fundamental to the process of building mature communities" (Montgomery, 1995, p. 251).

While social work students in Italy might have once perceived the English requirement of their degree course a mere formality, the current migrant situation in Europe has led to an urgent need for active focus on English in context. Specialised corpora can help reveal what language and discourse is 'appropriate' or at least recurring. In an ESP framework, corpus linguistics has mainly been used to focus on the behaviour of lexical items more or less in isolation, or at least only as regards collocational patterns. Boulton (2012) suggested that most researchers focused on smaller amounts of phraseology and collocations rather than discursive resources across the text. A more holistic approach focuses on discourse as it appears in the contexts of the different genres in a corpus.

This chapter shows how one such corpus was developed for an ESP course with social work students in an Italian university. First, a needs analysis of these students is presented, and then genres appropriate to social work are identified. Following this, the contents of the Social Work Corpus are described, including a detailed exploration of how one sub-corpus could be exploited for classroom use. Finally, suggestions for further development are made.

2. Needs analysis

Degrees in social work at Italian universities were aligned with European standards after the Bologna Declaration (1999), though there is often little compulsory social work theory and practice content (Frost, Höjer, & Campanini, 2013). Social work placement, often shadowing the staff of local authorities, is an important part of the course and social work graduates in Italy need to apply their university education to interpreting situations and intervening within their specific cultural and ethnic contexts (Campanini & Frost, 2004). We should note here that the current refugee crisis has led to some major changes in an Italian social worker's mandate. Italy is often the first point of arrival for refugees fleeing from conflict areas, with a huge increase in the number of unaccompanied minors, as well as incoming migrant workers, particularly from African nations. Figures for 2015-16 (Menonna, 2016) show the main countries of origin to be Eritrea, Nigeria, Gambia and Somalia, with a sharp rise in numbers from Bangladesh: all countries where English is the main foreign language. Besides these new concerns, ongoing social work in Italy deals increasingly with the additional issues of second-generation immigrants (Simone, 2016).

As regards English language competence, admission to the degree course requires incoming students to have a B1 level in the Common European Framework of Reference for languages (CEFR). Social work undergraduates – in this author's institution at least – tend to have a knowledge of general rather than specialised English and generally weak listening and production skills. While course attendance is not compulsory, students at Bologna University usually attend general English language classes prior to the ESP course with the aim of reaching a B2 level, focussing particularly on grammar and reading, since some of the core degree programme literature is delivered in English (Frost, Höjer, Campanini, & Kuhlberg, 2015). In their ESP course, therefore, students will need particular practice in speaking, listening and writing skills, while consolidating reading skills. Specialised terminology is best introduced in context, and language functions typical of social work also need to be presented and practised. The next section will discuss the importance of selecting appropriate learning material (Nesi, 2015).

3. Language and discourse for social work

Teachers need to evaluate the appropriateness of language and discourse, but this is challenging if the field is unfamiliar to them. The ideal informants are the social work practitioners themselves, given their familiarity with the required language use in social work practice, which the literature tends to discuss under the umbrella term 'communication'. Communication is mainly described here as synonymous with spoken interaction, which has always provided the basis for successful social work (e.g. Clark, 2000; Cross, 1974; Day, 1972; Pierson & Thomas, 2000; Thompson, 2010; Trevithick et al., 2004).

It is no surprise then that much literature in social work studies has focussed on the communication skills that the social work student needs to acquire (e.g. Lishman, 2009; Richards et al., 2005; Shulman, 2006; Thompson, 2010; Trevithick, 2005; Trevithick et al., 2004; Woodcock Ross, 2011), although there is little consensus as to what exactly constitutes appropriate communication (Richards et al., 2005; Trevithick et al., 2004) and where good models might be found. To be relevant to our purpose here, we shall focus solely on references to communication as spoken or written discourse, leaving aside non-verbal aspects such as body language and dress. We will divide references from social work literature into two types: one concerning general language references and the other genre-based references.

3.1. General language references

Social workers require both production and reception skills: "all workers need to develop appropriate communication skills both for face-to-face and for written communications" (Pierson & Thomas, 2000, p. 95). Oral skills are of fundamental importance in social work (Juhila, Mäkitalo, & Noordegraaf, 2014b), since "social workers spend more time in interviewing than in any other single activity. It is the most important, most frequently employed, social work skill" (Kadushin, 1990, p. 3).

Issues of register-awareness and the connotations of language emerge from practitioners' recommendations. For example, social workers need to "consider the context in which they are required to speak and to write, and to ensure that they develop a style that is appropriate and relevant for their audience" (Pierson & Thomas, 2000, p. 95), since clients are "not best helped by workers who use obscure, inaccurate, deceptive or demeaning language" (Clark, 2000, p. 181).

The pragmatic function of discourse is especially important, since "[p]rofessionals are less effective on their clients' behalf if they cannot communicate precisely and persuasively" (Clark, 2000, p. 181). Lishman (2009) lists the many different fields in which a social worker has to operate while giving an idea of the pragmatic functions of communication with which students need to be familiar:

> "Effective communication [...] includes, for example, providing basic care, giving advice, making assessments, providing care packages, counselling, writing reports, acting as advocates for service users, and working in interdisciplinary settings with health, education, housing and criminal justice" (p. 1).

This citation gives examples of both spoken and written discourse, as well as contextualising reading, writing, speaking, and listening skills in the professional life of a social worker. The importance of written communication skills in social work cannot be denied (Trevithick et al., 2004, p. 19). Lishman (2009) lists written material which may be required of the social worker, including letters, reports, and records for service users, agencies, and social workers.

Though the above authors were referring to a native-speaker environment, for the purpose of this chapter we may easily extend these examples to cover non-native contexts.

3.2. Genre-based references

A number of genres are described as particularly relevant to practitioners. Though dialogues between service users and social workers are important, they inevitably refer to, draw on, and follow other background material, since "all kinds of written reports and interprofessional meetings support that talk to complete a trajectory" (Juhila et al., 2014b, p. 9).

Such written reports include case recordings, defined as "brief narrative accounts of contacts with and about clients, summaries, and plans of work covering a period of time" (Payne, 2000, p. 44), case studies (describing the issues at stake), supervisor commentaries, and case narratives, in which the social worker notes down his/her thoughts during the interview and describes the rationale behind the intervention. In relation to the refugee crisis, being able to read and compile material intended for cross-border use would be a major advantage for Italian social workers.

Both Shulman (2006) and O'Hagan (1996) mention different fields with which competent social workers need to be familiar, since "the knowledge underpinning social work practice derives from many different sources" (O'Hagan, 1996, p. 8) and students must "become attuned to the ways in which society, government, courts and professional associations influence practice" (Shulman, 2006, p. xxi). They should keep up with legislative developments such as revisions to ministerial responsibilities with repercussions on health and social care at national and local level, both at the source – through dedicated governmental websites – and trickling down through the printed media to the service user. In particular, students need an awareness of:

> "law, social policy, philosophy (ethics), sociology, social administration, organisational policies, procedures and guidelines, numerous theories, [and] differing social work methods" (O'Hagan, 1996, p. 8).

Thus we may interpret these recommendations as requiring access to relevant material, including legislation and best practice recommendations produced by national and local governing bodies with relevance to social care, legal documents, as well as healthcare and medically related material. In an Italian context, this would provide material for contrast with students' own national situation, as well as raising awareness about other international applications.

Some practitioners (e.g. Pierson & Thomas, 2000) also draw attention to the need for familiarity with the academic genre, particularly the research article. This is due to increasing emphasis on a research culture in social welfare, in which "workers need to be able to communicate clearly with funders, research colleagues and research participants in order to produce high quality results and be able to disseminate their findings clearly and imaginatively in order to improve practice" (Pierson & Thomas, 2000, p. 95).

Information is highly interconnected in a number of specialised fields (c.f. ESP and Law, in Breeze, 2015) and social work is no exception. ESP students need to be aware of these connections and the dissemination of information across different genres. News articles constitute useful sources of information for the social worker as regards public reception and perception of social work legislation. News articles are a potential means to encourage students to have a proactive approach to their profession. Brawley (1997), for example, uses newspaper discourse to promote media advocacy among social worker students, "with a view to correcting distorted messages about vulnerable people given by the press, developing policies and services, and relaying important messages to large target audiences" (Trevithick et al., 2004, p. 30).

Social work students should therefore cultivate awareness of a wide range of genres, including reports and examples of social care and social work interaction, editorials and news reports, academic and legal genres, and government documents. While non-native speakers in an Italian context would certainly need to be able to read and compare such material with similar texts in Italian, current social work in Italy may also involve actual production of such material in English.

To sum up, communication as described by social work practitioners may be broken down in linguistic terms as being genre-based and macro skills-based, including familiarity with particular genres and relative genre constraints such as register, lexis and phraseology, the need for pragmatic awareness, and practice in all four macro skills, particularly those hitherto under-developed speaking skills. In particular, students must become aware of the linguistic nuances of

Chapter 8

spoken language in social work interviews in English in order to communicate successfully in the various interactions at the heart of social workers' professional practices. It is worth noting here that the students' university curriculum does not include active focus on the language used in their first language. Required reading prior to placement (Zini & Miodini, 2004) does feature a section on 'language', but only as regards what to talk about rather than how to say it and why. The focus of this chapter, therefore may be considered doubly useful for Italian social work undergraduates, since it applies to social work universally and is particularly important in a context where English is not the students' first language.

Given the connections between the different aspects of a social worker's profession, the sources of interest to language teachers should be considered as a network of resources. Once collated into an electronic corpus, authentic texts constitute relevant and topical material on which to base a meaningful language course for ESP Social Work undergraduates, following an integrated language-teaching approach whereby the four macro skills are taught in conjunction with each other. Such a corpus is described in the next section.

4. Introducing SocWoC

The Social Work Corpus (SocWoC) is intended as a repository of teaching material in electronic form. It has high 'face-validity' (Flowerdew, 1993, p. 239), including all the material the students are exposed to during their ESP course and featuring examples of many of the different genres the students may encounter in their future profession.

Divided into various sub-corpora, SocWoC is a *monitor corpus* in that it may constantly be added to[2], either vertically (by adding additional texts of the same genre) or horizontally (by adding new genres). It may thus also be updated to reflect new legislation or topics in the current interest. While its present

---

2. Though SocWoC was designed by the author, compiling a monitor corpus lends itself to collaborative effort and any additional material contributing to enhance the utility of a social work corpus would be much appreciated, in exchange, naturally, for access to the whole corpus.

composition, on which this chapter is based, is designed to give a 'snapshot' of current practice, thus consisting of documents published during the period 2015-16, there are plans to add material dating from earlier years to each sub-corpus to facilitate broader linguistic analysis. Though mainly consisting of material from UK sources 'written to be read', it also features transcripts of spoken material, including dialogues from social workers in a US context to allow any differences resulting from language variety to emerge. A brief description of each sub-corpus is given in Table 1.

Table 1. List of sub-corpora within SocWoC

Sub-corpus	Description	Number of words
MAT	training materials	75,237
ACAD	academic papers	172,011
NEWS	newspaper articles	156,441
GOV	government guidelines	80,692
GLO	glossary of social work terms	15,534
TOTAL		499,915

With the exception of most of the MAT texts, the corpus material was downloaded from the internet and stored in separate files without incorporating metadata. Material for the MAT sub-corpus was difficult to access since social work does not have much tradition of being observed and recorded and social work interviews are confidential encounters. For this reason, selected social work practice books (Bisman, 2014; Woodcock Ross, 2011) were used as well as training material available online from reliable sources such as the SCIE website (www.scie.org).

The contents of the five sub-corpora in their current state are described as follows:

- MAT contains training material intended for classroom use by social work students in both the UK and the US in the form of case studies, case narratives, supervisor commentaries, and the transcriptions of social work interviews.

- ACAD contains academic papers from the field of social work. It currently features some 30 articles published in the *British Journal of Social Work*, representing the British Association of Social Workers.

- NEWS features articles appearing in the Guardian and the Daily Telegraph: quality UK newspapers from both sides of the political spectrum. Texts containing the node words *social work** were collected.

- GOV contains a selection of government-sponsored examples of good practice in social care available online. Given SocWoC's predominantly British focus, at least in this initial stage, sources included the UK's Department for Education, the Department of Health, and government-funded institutions such as the Social Care Institute for Excellence.

- GLO contains social work terms with their corresponding definitions from glossaries in social work textbooks and websites, such as local government sites, and the online resources of institutions offering social work degree courses.

Metadata was included to associate each dialogue with the corresponding case studies and case narratives, while manual markups of the beginning and end of both social worker and service user turns was added to enable semi-automatic extraction of the discourse of the different categories of speaker. In the next section, we discuss possible uses for SocWoC by describing some initial corpus linguistic findings from one sub-corpus and giving some brief suggestions of how these might be exploited in the ESP social services classroom. We shall conclude by considering the corpus as a whole.

5. Exploiting SocWoC

Without the knowledge of or ability to practise effective communication, "the social worker is unlikely to use encounters with users of services or colleagues for purposeful communication" (Lishman, 2009, p. 207). This is particularly true for speakers of other languages. Thus the teacher needs to focus both on awareness-raising activities and on language practice. In order to identify important language features, the teacher can use corpus linguistics techniques to query SocWoC in order to prepare teaching materials. Corpus linguistics is certainly nothing new in the field of ESP (Biber, Reppen, & Friginal, 2010, p. 559). It has been exploited in an ESP and English for Academic Purposes (EAP) framework (e.g. Boulton, Carter-Thomas, & Rowley-Jolivet, 2012) particularly for investigating genre and vocabulary. Examples of research into the possibilities offered by these techniques in relation to language teaching include Krishnamurthy and Kosem (2007), Hyland and Tse (2007), Ghadessy, Henry, and Roseberry (2001), Scott and Tribble (2006), O'Keeffe, McCarthy, and Carter (2007), Breeze (2015), and Boulton (2016).

In what follows I provide suggestions for teachers to exploit SocWoC with students, although it is of course also possible to give learners themselves direct access to the data in line with 'data-driven learning' (e.g. Johns, 1991; Boulton, 2016). Given space restrictions, I will focus here on an analysis of the MAT sub-corpus alone, more specifically the part containing the transcripts of social work interviews.

5.1. Mining the MAT sub-corpus for classroom use

Preliminary corpus investigation highlighted a variety of elements meriting further attention in the classroom. The examples given below are mainly concerned with pragmatics, such as preferred choice of process according to speaker turn, the use of interrogatives, the discussion of 'feelings' in what are often extremely distressing situations, the importance of figurative language and paraphrasing, as well as issues regarding language variety. A final example is a semantic analysis.

Since the dialogues in MAT are tagged for speaker turns, wordlists may be run for either speaker. Social worker turns, for example, contain a predominance of **mental processes** (*want, need, like, think*):

> (1) I am glad you are sharing this with me. You have told me in previous sessions that at various times in your life you have had these feelings. What do you think is going on now that you are considering killing yourself?
>
> (2) What I want to do, first of all, is explain to you why we are getting together, if you like. [...] We need to do an assessment today, aspects of your life and things that might affect your offending behaviour that we can work on together. Do you know anything about the Youth Offending Team?

These mental processes often appear in **interrogative forms** as in Examples 1 and 2. Interrogatives are used not only to establish facts (e.g. *Do you know anything about the Youth Offending Team?*), but also to help the social worker to "put feelings into words" and "reach for feelings": an important part of the social work interview (Shulman, 2006, p. 136), as in Example 1 (*What do you think is going on now that you are considering killing yourself?*). The different functions of interrogatives in the social work interview lend themselves to class discussion, while question types and their functions (e.g. Richards et al., 2005; Trevithick, 2005) can be further investigated in the classroom using SocWoC.

Exploration of the wordlists enables identification of other phrases used to **'reach for feelings'**. Some of these are triggered by the lexical item 'feel':

> (3) <SERVICEUSER> A fog is all around me. Nothing matters, I don't feel anything.
>
> <SOCIALWORKER> How do you feel about having this fog?
>
> <SERVICEUSER> I don't feel anything. I never thought about it. [...] I feel like I am in the center of a cloud.

Example 3 illustrates the utility of **figurative language** for discussing feelings in social work talk.

Searching the corpus for *like* reveals the same function fulfilled by phrases such as *sounds like, is it like,* as in Example 4:

> (4) It sounds like it's quite painful ... and that when you're walking, checking over your shoulder, is it like you're running away from something?

Classroom exploitation could focus on classifying the pragmatic functions of such figurative language in both speaker turns. The frequency of mental processes used in the interrogative form, interrogatives generally, and phrases acting as explanations all relate to facework and politeness (Goffman, 1955), which are central issues in social work interviews (Juhila et al., 2014b).

Wordlists show a high frequency of reporting verbs, whose different functions in social work have been described by Juhila, Jokinen, and Saario (2014a). Reported speech also plays an important part in **paraphrasing**, a fundamental communication skill (Shulman, 2006; Trevithick, 2005). An example is shown in Example 5:

> (5) Just take a minute and think about how you are feeling. You have said before that you have trouble being aware of and articulating your feelings.

In Example 5 the service user's words (signalled by '*you have said before*') are paraphrased by the social worker. This enables him/her to refer to the service user's problems with discussing feelings, and thus move towards a possible solution. In class, students could look at examples of paraphrasing marked by reporting speech and practise transforming the service user's words in different contexts.

Since both UK and US English is represented in the material in MAT, differences between these two **language varieties** may be highlighted. A frequent phrase in the British dialogues was *at the moment*. While this phrase was not found in the US social worker turns, the same meaning emerged in the frequent phrase *right now*. Similarly differences in verb colligation emerged: *talk with* versus *talk to*, for example. Such investigations in the classroom make it possible to promote awareness-raising as regards language variety.

Finally, an investigation into the **key semantic fields** in the dialogues provides opportunities for focussing on ESP phraseology. Using WMatrix (Rayson, 2008), the dialogues were compared with the spoken section of the British National Corpus. One of the top semantic domains was 'worry'. While this in itself is not unexpected, given the nature of most social work interviews, intuition alone might not be able to supply all the vocabulary related to this field (examples in Table 2).

Table 2. Concordances exemplifying the semantic domain 'worry'

I act like I don't	**care.**	But then when I 'm in
mom to hit you. We have serious	**concerns**	about your bruises
you have said before that you have	**trouble**	being aware of and
Doug, you seem a little	**anxious**	about me being here.
Graeme had presented a lot of mental	**distress**	, and had been hungry, cold
Yoko is under a lot of	**stress**	right now
That list of things that	**bother**	you included feeling unloved

Concordances from this semantic domain may be used to teach a range of sometimes very sensitive vocabulary choices in context, including collocations (*serious concerns*), colligation (*anxious about*), specialist combinations (*present + mental distress*) and phraseology (*under* + [quantity] + *stress*). These are just some examples of the types of language items that could merit further linguistic investigation by/for students in one sub-corpus of SocWoC. The final sub-section will examine discourse in context across the different parts of SocWoC.

5.2. Classroom focus across SocWoC

The social work profession is people-focussed. One way of identifying key participants is to compare wordlists from each sub-corpus with a general reference corpus such as the BNC:

- MAT: *client, supervisor, families, parents, children, inpatient*

- ACAD: *children, students, participants,* [social] *workers, carer, family*

- GOV: *staff, adult, child, people, person, providers, carers, professionals, inpatients*

- NEWS: *social workers, child/children, people, families, refugees, staff, carers, practitioners*

- GLO: *person, people, carer/s, customer, adults, organisations, user/s, worker/s, families, authorities, children*

One activity could be to investigate the differences and similarities between the items in these lists. Why do we find refugees only in one part of SocWoC? Which participants appear in all parts of SocWoC? Investigation could continue into a comparison of the items, thus allowing focus on, for example, preferred or excluded terms in particular genres. Corpus tools such as SketchEngine (Kilgarriff, Rychly, Smrz, & Tugwell, 2004) may be used to generate a word profile to focus on the search word in context. The most frequent processes associated with *children,* for example, were found to be different across the sub-corpora, as was the positioning of children as subject or object. This sort of investigation is particularly important for non-UK based social work students, since the context where these students expect to practise may be different from the UK context. In the family-oriented Italian system, families are the main actors in solving people's problems whereas the state plays a subordinate role (Frost et al., 2013, p. 341). This is slightly different to the more individualistic cultural models in the UK and US and students must be aware of how people's needs are dealt with in the different contexts.

We can, of course also look outside the corpus to gain further information regarding, for example, current problematic issues. With reference to the investigation of *children* outlined above, we might decide to follow the trajectory of children's social work through the discourse of different genres in relation to legislative items such as the recent Children and Social Work Bill and its implications for social work in the UK, including both attitudes to and receptions of such measures.

6. Conclusion

Italian social work undergraduates were given little theory and practice as regards the social work interview in their own first language, so that focus on the social work interview during lessons was more than just 'language comparison' but actually became a way to teach the functions of language in this context. Thus, we might consider that the way forward for the ESP teacher is increasingly moving towards combining language with content teaching, in order to be able to fully exploit the nuances and functions of the language. This requires the ESP teacher to be ever more prepared. In this chapter, we have described how a specialised corpus can be created and can inform the development of material for use in the classroom activities of an ESP course. The chapter focussed on the needs of a particular subset of ESP students, and suggested that information from social work literature could be used to guide the compilation of a suitable corpus of classroom material. In line with the student-centred approach of task-based language learning (e.g. Nunan, 2004), corpus linguistics techniques may then be applied in order to exploit this material in the classroom.

Further development and subsequent exploitation of SocWoC will need to take into account other factors important for the Italian social work context, particularly the need for texts to be appropriate for present-day social work in Italy, given the recent influx of immigrants using English as a lingua franca (Simone, 2016). Future work will thus involve refining the needs analysis, for example through pre-service and stakeholder surveys (e.g. Chovancová, 2014). Further development would also require fully integrating the findings briefly described here within the design of a comprehensive course for social work students, while it would also be important to be able to assess outcomes in terms of language awareness in students on completing such a course.

References

Biber, D., Reppen, R., & Friginal, E. (2010). Research in corpus linguistics. In R. Kaplan (Ed.), *The Oxford Handbook of Applied Linguistics* (pp. 548-567). Oxford University Press. https://doi.org/10.1093/oxfordhb/9780195384253.013.0038

Bisman, C. (2014). *Social work: value-guided practice for a global society.* Columbia University. https://doi.org/10.7312/bism15982

Boulton, A. (2012). Corpus consultation for ESP: a review of empirical research. In A. Boulton, S. Carter-Thomas, & E. Rowley-Jolivet (Eds), *Corpus-informed research and learning in ESP: issues and applications* (pp. 263-293). John Benjamins. https://doi.org/10.1075/scl.52.11bou

Boulton, A. (2016). Integrating corpus tools and techniques in ESP courses. *ASp, 69*, 113-137. https://doi.org/10.4000/asp.4826

Boulton, A., Carter-Thomas, S., & Rowley-Jolivet, E. (Eds). (2012). *Corpus-informed research and learning in ESP: issues and applications.* John Benjamins. https://doi.org/10.1075/scl.52

Brawley, E. A. (1997). Teaching social work students to use advocacy skills through the mass media. *Journal of Social Work Education, 33*(3), 445-60.

Breeze, R. (2015). Teaching the vocabulary of legal documents: a corpus-driven approach. *ESP Today, 3*(1), 44-63.

Campanini, A. M., & Frost, E. (Eds). (2004). *Social work in Europe. Commonalities and differences.* Carocci.

Chovancová, B. (2014). Needs analysis and ESP course design: self-perception of language needs among pre-service students. *Studies in Logic, Grammar and Rhetoric, 38*(1), 43-57. https://doi.org/10.2478/slgr-2014-0031

Clark, C. (2000). The use of language. In M. Davies (Ed.), *Blackwell encyclopaedia of social work* (pp. 181-183). Blackwell Publishers Ltd.

Cross, C. P. (Ed.). (1974). *Interviewing and communication in social work.* Routledge.

Day, P. D. (1972). *Communication in social work.* Pergamon Press.

Flowerdew, J. (1993). Concordancing as a tool in course design. *System, 21*, 231-244. https://doi.org/10.1016/0346-251X(93)90044-H

Frost, E., Höjer, S., & Campanini, A. M. (2013). Readiness for practice: social work students' perspectives in England, Italy, and Sweden. *European Journal of Social Work, 16*(3), 327-343. https://doi.org/10.1080/13691457.2012.716397

Frost, E., Höjer, S., Campanini, A. M., & Kuhlberg, K. (2015). Outsiders and learners: negotiating meaning in comparative European social work research practice. *Qualitative Social Work, 16*(4), 465-480. https://doi.org/10.1177/1473325015621124

Ghadessy, M., Henry, A., & Roseberry, R. L. (Eds). (2001). *Small corpus studies and ELT: theory and practice*. John Benjamins. https://doi.org/10.1075/scl.5

Goffman, E. (1955). On face-work: an analysis of ritual elements of social interaction. *Psychiatry: Journal for the Study of Interpersonal Processes, 18*(3), 213-231. https://doi.org/10.1080/00332747.1955.11023008

Hyland, K., & Tse, P. (2007). Is there an "academic vocabulary"? *TESOL Quarterly, 41*(2), 235-253. https://doi.org/10.1002/j.1545-7249.2007.tb00058.x

Johns, T. (1991). Should you be persuaded: two samples of data-driven learning materials. *Classroom concordancing: ELR Journal, 4*, 1-16.

Juhila, K., Jokinen, A., & Saario, S. (2014a). Reported speech. In C. Hall, K. Juhila, M. Matarese & C. van Nijnatten (Eds), *Analysing social work communication. Discourse in practice* (pp. 154-172). Routledge.

Juhila, K., Mäkitalo, Å., & Noordegraaf, M. (2014b). Analysing social work interaction. Premises and approaches. In C. Hall, K. Juhila, M. Matarese & C. van Nijnatten (Eds), *Analysing social work communication. Discourse in practice* (pp. 9-24). Routledge.

Kadushin, A. (1990). *The social work interview* (3rd ed.). Columbia University Press.

Kilgarriff, A., Rychly, P., Smrz, P., & Tugwell, D. (2004). The Sketch Engine. In *Proceedings EURALEX 2004, Lorient, France* (pp. 105-116).

Kornbeck, J. (Ed.). (2003). *Language teaching in the social work curriculum*. Logophon Verlag.

Kornbeck, J. (2008). A "reverse mission" perspective on second-language classes as part of social work education programmes. *Portularia: Revista de Trabajo Social, 8*(2), 247-264.

Krishnamurthy, R., & Kosem, I. (2007). Issues in creating a corpus for EAP pedagogy and research. *Journal of English for Academic Purposes, 6*, 356-373. https://doi.org/10.1016/j.jeap.2007.09.003

Lishman, J. (2009). *Communication in social work*. (2nd ed.). Palgrave Macmillan.

Menonna, A. (2016). *L'Immigrazione straniera in Italia: tendenze recenti e prospettive*. Fondazione ISMU. Settore Monitoraggio.

Montgomery, M. (1995). *An introduction to language and society*. Routledge.

Nesi, H. (2015). ESP corpus construction: a plea for a needs-driven approach. *ASp, 68*, 7-24. https://doi.org/10.4000/asp.4682

Nunan, D. (2004). *Task-based language Teaching*. Cambridge: Cambridge University Press. https://doi.org/10.1017/CBO9780511667336

O'Hagan, K. (1996). *Competence in social work practice: a practical guide for professionals*. Jessica Kingsley Publishers.

O'Keeffe, A., McCarthy, M., & Carter, R. (2007). *From corpus to classroom*. Cambridge University Press. https://doi.org/10.1017/CBO9780511497650

Payne, M. (2000). Case recording. In M. Davies (Ed.), *Blackwell encyclopaedia of social work* (p. 44). Blackwell Publishers Ltd.

Pierson, J., & Thomas, M. (2000). *Collins dictionary of social work*. Collins.

Rayson, P. (2008). From key words to key semantic domains. *International Journal of Corpus Linguistics, 13*(4), 519-549. https://doi.org/10.1075/ijcl.13.4.06ray

Richards, S., Ruch, G., & Trevithick, P. (2005). Communication skills training for practice: the ethical dilemma for social work education. *Social Work Education, 24*(4), 409-22. https://doi.org/10.1080/02615470500096928

Scott, M., & Tribble, C. (2006). *Textual patterns: keywords and corpus analysis in language education*. John Benjamins. https://doi.org/10.1075/scl.22

Shulman, L. (2006). (5th Ed.), *The skills of helping individuals, families, groups, and communities*. Thomson Brooks/Cole.

Simone, D. (2016). Servizio sociale e immigrazione. In A. M. Campanini (Ed.), *Gli Ambiti di Intervento del Servizio Sociale* (pp. 205-224). Carocci Faber.

Thompson, N. (2010). *Theorizing social work practice*. Palgrave Macmillan.

Trevithick, P. (2005). *Social work skills: a practice handbook*. Open University Press.

Trevithick, P., Richards, S., Ruch, G., Moss, B., Lines, L., & Manor, O. (2004). *SCIE knowledge review 6: teaching and learning communication skills in social work education*. Social Care Institute for Excellence.

Woodcock Ross, J. (2011). *Specialist communication skills for social workers: focusing on service users' needs*. Palgrave Macmillan.

Zini, M. T., & Miodini, S. (2004). *Il colloquio di aiuto. Teoria e pratica nel servizio sociale*. Carocci Faber.

9 Identifying and responding to learner needs at the medical faculty: the use of audio-visual specialised fiction (FASP)

Rebecca Franklin-Landi[1]

Abstract

Since the development of Task-Based Language Teaching (TBLT) in the 1980's, learner needs have been central to English for Specific Purposes (ESP) teaching and learning, including in the field of English for Medical Purposes (EMP). This paper reports on two studies, conducted at Nice University Medical Faculty between October 2015 and March 2016, designed to analyse and respond to learner needs in EMP. While the first study was a needs analysis of medical students, the second one concentrated on certain needs previously identified and sought to satisfy them using audio-visual specialised fiction or 'Fiction À Substrat Professionnel' (FASP). It focusses on the use of a clip from a medical television series and how it was used in the classroom to reinforce good medical practice through the identification of on-screen procedural problems. Qualitative data were collected using questionnaires and interviews and data analysis showed an evolution in students' critical analysis and in their cultural and medical practice awareness. The study therefore suggests that it is possible to satisfy a demand for quality language education with students who are not language specialists and that audio-visual FASP seems to be an interesting and useful pedagogical tool in ESP to meet the differing needs of specific professions.

Keywords: task-based language teaching, specialised fiction, FASP, EMP, learner needs, sociocultural approach.

---

1. Nice University, Nice, France; rebecca.landi@unice.fr

How to cite this chapter: Franklin-Landi, R. (2017). Identifying and responding to learner needs at the medical faculty: the use of audio-visual specialised fiction (FASP). In C. Sarré & S. Whyte (Eds), *New developments in ESP teaching and learning research* (pp. 153-170). Research-publishing.net. https://doi.org/10.14705/rpnet.2017.cssw2017.750

Chapter 9

1. Introduction

EMP has been a flourishing field of research in English-speaking academia for several decades since Maher's (1986) article clearly set out the state of the art in this speciality. Within the French academic world, specific EMP research became popular during the late 1980's and 1990's in parallel with the development of ESP. As ESP finds itself at the intersection of several disciplines, researchers tend to concentrate on particular areas dependent on their own academic interests or backgrounds (Sarré & Whyte, 2016) and this heterogeneity is equally found in EMP. It is for this reason that different approaches within ESP/EMP research have developed (Gledhill & Kübler, 2016; Sarré & Whyte, 2016) didactics and classroom practice, concept-oriented linguistics, and context-oriented linguistics. It is important, however, to emphasise the fact that these differing research directions do not function independently and inter-relations exist between them.

In the field of didactics and classroom practice in particular, learner needs have been central to ESP teaching and learning since the 1980's, as it is usually the starting point of ESP course design (Hutchinson & Waters, 1987). This chapter will contribute to this area by reporting on two studies related to EMP needs, including a pedagogical intervention using audio-visual specialised fiction (FASP) designed to develop EMP students' cultural knowledge. This contribution therefore has two main objectives: (1) to identify the needs of EMP learners at a French medical faculty; and (2) to evaluate to what extent the use of audio-visual specialised fiction can meet these needs. After reviewing the relevant literature, we will present the methods used in both studies and the results obtained, before discussing how these results compare with similar studies.

2. Literature review

Applied linguistics, described as "the theory and practice of language acquisition and use" (Kramsch, 2000, p. 317), was perhaps the first attempt by academic researchers to apply language theory to real-life situations. It forms the basis for many of the pedagogical and language-acquisition theories that have developed over the years. The two principal applied-linguistic approaches to ESP are the concept-oriented approach and the context-oriented approach (Gledhill & Kübler, 2016). The concept-oriented approach concentrates on the technical language and linguistic structures employed within a specialist domain, for example through metaphor analysis (English, 1998), whereas the context-oriented approach considers the social or cultural context when analysing specialist language. Within the context-oriented approach to ESP, we find the specific branch of discourse analysis, which concentrates on the vocabulary, grammar, and sentence construction used in ESP texts. Discourse analysis can, among other pedagogical uses, provide the basis for glossaries, medical dictionaries, and vocabulary lists, as suggested in a study on deceptive cognates in English and Spanish medical articles (Divasson & León, 2002).

The third main branch of ESP research, that is didactics, developed with the evolution of TBLT in the 1980's and 1990's (Ellis, 1994, 2003, 2013; Foster, 1999) where classroom practice became the subject of academic research. TBLT has been widely put into use in the EMP classroom (Faure, 2003; Pavel, 2014) and is still being researched today. Indeed, task-based language teaching is at the root of a communicative approach to EMP in which role-play and simulations of specific medical situations are employed to develop the learner's oral professional autonomy (Hoekje, 2012). Didactics has often, however, been considered as being the "poor relation" to the two more linguistically-based research specialities (Sarré & Whyte, 2016, pp. 150-151). Concentrating more closely on what is actually done in a language classroom and on the concrete methods and techniques that teachers employ to help their students acquire the target language, didactic research is accused of being too context-specific to be easily replicated and for the pedagogical practices reported on to be adopted on a wider scale (Whyte & Sarré, 2016). This negative image has recently been called into question by Sarré and Whyte (2016), who demand that a clear research framework for ESP didactics be established to avoid an anecdotal style in academic articles and to enable this research to be recognised at its true value. Despite, or perhaps because of, this tendency to discount practical teaching analysis, attempts have been made to analyse the different currents present in didactic research.

Chapter 9

In ESP pedagogy, Belcher (2004) clearly defines the three main trends in teaching present in the mid-2000's and which continue to this day: the sociodiscoursal, sociocultural, and sociopolitical approach. The sociocultural approach is particularly favoured in EMP due to the specifically cultural aspect of medicine and healthcare in general, which Mourlhon-Dallies (2008) emphasises when she speaks of the intercultural dimension present when teaching a language for medical usage, as "medicine is about the body, religion, and death"[2] (p. 170, our translation). When teaching medical English, other non-linguistic factors are in play and these sociocultural elements are what Hoekje (2012) points to as the future of EMP research when she writes that "the cultural basis of healthcare practice needs further recognition" (p. 2). This gap in the research between what EMP academic specialists investigate and the expectations concerning English that medical professionals actually hold has also been identified in France (Carnet & Charpy, 2017). It is for this reason that alternative teaching methods have developed which aim not only to meet specific linguistic needs, but also to provide the second language learner with a certain cultural knowledge of the specific domain studied. One of these methods, which will provide the basis for this research study, is the use of specialised fiction which has been termed '*fiction à substrat professionnel*', or FASP, in the French literature.

The term FASP first appeared in an article by Petit (1999) where he identifies the specific genre of professional literary fiction and its three principle aspects: (1) the novels are international bestsellers, (2) the authors repeat their success with several bestselling books in the same genre, and (3) the authors are specialists within the particular discipline that they write about. The third characteristic is especially important as it explains the specific professional language, discourse, and vocabulary found in these works, and of course adds to its authenticity. As Isani (2004b) explains, "the profession in question is the very pivot of plot and character dynamics" (p. 26). The fact that a character is, for example, a doctor, is essential to the plot and is not just a background detail.

---

2. In Mourlhon-Dallies's (2008) own words: "la médecine touche au corps, à la religion, à la mort. La dimension interculturelle est donc également très présente dans l'exercice des métiers en question" (p. 170).

In France, FASP has proved to be a very popular subject of research in ESP pedagogy, especially in the domain of English for Legal Purposes (ELP) (Chapon, 2015; Genty, 2010; Isani, 2004a, 2006, 2011; Villez, 2004;). After Petit's (1999) initial definition of professional literary fiction within the field of ESP, the definition was quickly expanded upon to include movies (Isani, 2004a) and television series (Villez, 2004) anchored in a professional milieu. Indeed, what has come to be known as audio-visual FASP continues to provide an interesting pedagogical tool in today's ELP classroom as shown by Chapon's (2015) recent doctoral thesis on the subject, where she proved the utility of audio-visual FASP in teaching the adversarial judicial system of Common Law in English-speaking countries. In the field of EMP, Charpy (2004, 2005, 2010, 2011) has worked extensively on literary FASP, but the use of audio-visual FASP has been little studied (with the exception of Carnet's (2015) research using the series *House MD*). There is still room, therefore, for development and innovation within didactic research in EMP and, more particularly, in the use of audio-visual FASP. The two studies whose methods are presented in the next section aim to contribute to bridging this gap.

3. Methods

3.1. Pilot study

Study 1 was a pilot study analysing the needs of medical students carried out in October 2015 on 54 second-year (L2) students and on 49 third-year (L3) students using qualitative data collection[3]. A short multiple-choice questionnaire[4] on students' general attitudes and needs in EMP was given to the students, and the same questionnaire was also used with ten medical professors. Seven professors answered the questions independently, and three of them answered them during an interview with the researchers, thus allowing more expert analysis and

---

3. The first year of medical studies in France (the PACES) is open to anyone, with a selective examination at the end which allows only a very limited number of students to access the second year of medical studies. First-year students do not have English lessons and are not tested on their level at Nice Medical Faculty.

4. Supplement, part 1: https://research-publishing.box.com/s/lgnkjny733kallj5961qh75q8qlks60i

discussion of the specific needs in EMP from the professionals' point of view. A pen-and-paper version of the questionnaire was given to the L2 students during their first English lesson of the year and all of them responded (=100% response rate). The objective was to see what they thought their needs were, what type of lessons they wanted before starting English, and then asking students belonging to the year above (L3) what they thought of the lessons they had had and to see if those lessons had met their expectations and needs. The student needs were then compared to those identified by medical professionals.

3.2. Study 2

Study 2 aimed to fine-tune the questionnaire given in the pilot study and to inform the design of an audio-visual FASP sequence to meet the needs previously identified. This study was conducted on L3 medical students in March 2016 during two non-obligatory 60-minute classes. 59 (=100%) answered the opening needs-analysis questionnaire[5] and 38 of the initial 59 (=64%) took part in the FASP experimentation phase. Pre- and post-teaching questionnaires[6] were used to assess students' attitudes to FASP, as well as a content and language test to evaluate their understanding of the video sequence[7]. The pre-FASP questionnaire[8] was adapted from an American study (Czarny et al., 2008) given to the students to identify their general profile. After some introductory questions (five open-ended questions), there were more specific ones dealing with television medical shows (six multiple-choice questions) to discover if they watched this type of programme, what shows they watched, the frequency of viewing, and language(s) they watched in, etc. Finally, the last section to the questionnaire (three multiple-choice questions) aimed to identify their attitude towards medical shows, how accurate they found the portrayal of hygiene and good medical practice, and whether they discussed medical issues with their friends and family. Once this general questionnaire had been completed, the students watched a five minute

---

5. Supplement, part 2: https://research-publishing.box.com/s/lgnkjny733kallj5961qh75q8qlks60i

6. Supplement, parts 2 and 3: https://research-publishing.box.com/s/lgnkjny733kallj5961qh75q8qlks60i

7. Supplement, part 4: https://research-publishing.box.com/s/lgnkjny733kallj5961qh75q8qlks60i

8. Supplement, part 2: https://research-publishing.box.com/s/lgnkjny733kallj5961qh75q8qlks60i

video clip taken from Season 11, Episode 4 of *Grey's Anatomy*, which deals with a patient who becomes paralysed in his lower spine after a thoracic stent graft. The students watched the clip twice and completed the content and language test[9] at the same time. The clip was shown without prior explanation to evaluate what the students had effectively understood. After viewing the clip and completing the test individually, a class discussion was held. The students were introduced to the UK General Medical Council's (GMC) guide to good medical practice[10], which identifies four key domains: (1) knowledge, skills and performance, (2) safety and quality, (3) communication, partnership and teamwork, and (4) maintaining trust with the patient and with colleagues. These guidelines were used as they are internationally recognised, and because similar, ethically-oriented guidelines do not exist in the US, perhaps due to the healthcare system being viewed principally as a business. The video clip was then considered in the light of the GMC guidance and students discussed how the television show respected, or not, the principles related to each of the four domains. Finally, the students completed a post-FASP questionnaire[11] to find out how accurate they found the depiction of medical practice in the video clip after class discussion and to collect their personal reactions towards this teaching method.

4. Results

4.1. Pilot study

All three groups of participants (second- and third-year students, and medical professors) had to answer the same first three questions (Table 1). The first two questions assessed the participants' level of bias, their personal feelings towards English, and its usefulness for them individually. The third question was to determine what skills were most important for medical students. The questionnaires diverge from question four onwards: in L2 it concentrated on the

---

9. Supplement, part 4: https://research-publishing.box.com/s/lgnkjny733kallj5961qh75q8qlks60i

10. http://www.gmc-uk.org/guidance/good_medical_practice.asp

11. Supplement, part 3: https://research-publishing.box.com/s/lgnkjny733kallj5961qh75q8qlks60i

Chapter 9

type of pedagogy that they wanted for their future English lessons, whereas in L3 they were asked for their favourite and least favourite aspects of their English lessons. Meanwhile, the professors were asked what should be taught and their preferred pedagogy for medical English.

Table 1. Answers to the pilot study questionnaires

	L2 (54)		L3 (49)			Medical Professors (10)		
	No.	%		No.	%		No.	%
Question 1: For me, English is…								
Necessity	28	52	Necessity	26	53	Necessity	6	60
Investment	9	16.6	Investment	8	16.3	Investment	2	20
Effort	2	3.7	Effort	-	-	Effort	-	-
Pointless	-	-	Pointless	1	2.1	Pointless	-	-
Pleasure	13	24	Pleasure	13	26.5	Pleasure	1	10
Other	2	3.7	Other	1	2.1	Other	1	10
Question 2: I think I need English lessons as a medical student for…						**Question 2: I think medical students need English lessons for…**		
Foreign Patients	47	87	Foreign Patients	41	83	Foreign Patients	6	60
Med. Journals	49	90	Med. Journals	44	89	Med. Journals	10	100
Int. Conferences	46	85	Int. Conferences	44	89	Int. Conferences	10	100
Personal Life	36	66	Personal Life	33	67	Personal Life	5	50
No need	-	-	No need	1	2.1	No need	-	-
Question 3: For me, the most important aspect of English is…								
Reading	6	11	Reading	4	8	Reading	1	10
Writing	-	-	Writing	1	2.1	Writing	-	-
Oral interaction	40	74	Oral interaction	41	83.6	Oral interaction	6	60
Presentations	2	3.7	Presentations	1	2.1	Presentations	2	20
Listening	5	9.3	Listening	1	2.1	Listening	-	-
Grammar	-	-	Grammar	-	-	Grammar	-	-
Vocabulary	1	2	Vocabulary	1	2.1	Vocabulary	-	-
Other	-	-	Other	-	-	Other	1	10
Question 4: At the medical faculty I want my English lessons to be…			**Question 4a: At the medical faculty my favourite aspect of my English lessons is…**			**Question 4a: As a doctor, I think English lessons should concentrate on…**		
Trad. Lessons	5	6	Small groups	7	14.5	Oral work	6	60
Oral groups	27	31	Oral work	11	22.5	Med. vocab	-	-
Med. vocab	44	51	Med. vocab	12	24.5	Med. culture	3	30
Lectures	2	2	Med. culture	4	8	Med. ethics	1	10
New technology	9	10	Med. ethics	5	10	Med. articles	10	100
Other	-	-	Med. articles	4	8	Other	-	-
			Only ENG	6	12.5			
			Other	-	-			

		Question 4b: At the medical faculty my least favourite aspect of my English lessons is...			Question 4b: As a medical professor, I think medical faculty English should be taught...		
		Small groups	2	4	Only ENG	8	80
		Oral work	6	12.5	New technology	6	60
		Med. vocab	5	10	Tutors	7	70
		Med. culture	8	16.3	ENG lectures	2	20
		Med. ethics	2	4	FR lectures	-	-
		Med. articles	18	36.7	Grammar/ vocab	3	30
		Only ENG	2	4			
		Other	6	12.5			
Question 5: In my opinion a French doctor and an American doctor do fundamentally the same job.							
Agree	35	65	Agree	25	51	Agree	10	100
Disagree	5	9	Disagree	13	26.5	Disagree	-	-
Don't know	14	26	Don't know	11	22.5	Don't know	-	-

Three important trends emerge from these results: (1) a general acceptance of the necessity for English in medical studies, (2) the importance of oral skills, and (3) the desire for specific medical terminology expressed by the students yet not at all by practising doctors. In fact, all three medical professors who answered the questions in an interview (as opposed to the pen-and-paper questionnaire) talked about the importance of understanding the cultures and mentalities of English-speaking countries to be fully autonomous in the English language, with specific mention of the importance of humour and understanding cultural references. And yet, medical culture is the second least favourite aspect of English lessons for L3 students. This result could be explained by medical students' lack of experience – they feel that their inability to understand certain medical situations can be simply remedied by learning the correct vocabulary. This would explain the popularity of this choice in Question 3, as students do not yet realise the importance of cultural knowledge when dealing with foreigners, something that experienced doctors do.

4.2. Study 2

Because of the emphasis placed on the cultural aspect of language learning by the medical professionals, and the lack of enthusiasm for this cultural dimension from the L3 students, it was decided to try a FASP-based sequence with the students to combine language acquisition, an authentic cultural setting, and consideration of specific professional skills. A clip from *Grey's Anatomy* was chosen as it offered the opportunity to explore the question of good medical practice through illustrating various examples of bad practice[12]. This series was selected as it is very popular with students and easily available in English, therefore students can continue watching the series outside of class, giving them more exposure to medical language and encouraging their linguistic and cultural development. The students were asked to identify the problems in the clip, to discuss why these actions were problematic and give possible solutions. The comparison between the reality of the GMC guidelines, the fictional depiction of hospital practice, and their personal knowledge of the French medical system highlighted the differences between the medical cultures of the USA, the UK, and France, and encouraged them to develop their critical analysis of audio-visual FASP and to remember the internationally recognised pillars of good medical practice.

The didactic sequence began with a needs-analysis questionnaire, following on from the pilot study. Out of 59 students, 91.5% (n=54) claimed they watched television and 92.5% of these (n=50) said they had watched a medical drama in the past, with the most popular show proving to be *Grey's Anatomy* for 84% (n=42) of them and 48% (n=24) of medical drama viewers claiming that they watched this particular TV show at least once a week. The answers given to this first questionnaire therefore confirmed that most medical students at Nice University watch medical programmes and also proved the popularity of *Grey's Anatomy*.

The students then watched a scene from *Grey's Anatomy* and answered a content and language test[13] while watching to evaluate their comprehension of medical terminology in the clip. They were not only tested on what they had understood from a medical point of view, but also in terms of the storyline.

---

12. In the clip an operation begins before the patient's family gives their consent, there are occasions when doctors walk into the operating theatre during open-heart surgery without respecting proper hygiene protocol, and the doctors communicate more about their personal lives than about the surgery being performed.

13. Supplement, part 4: https://research-publishing.box.com/s/lgnkjny733kallj5961qh75q8qlks60i

Table 2. Content and language test: key and results

2) What have you understood about the patient's medical situation?	3) What have you understood about the doctors' personal lives?
• He has an aneurysm ✓ • He has had a thoracic stent graft ✓ • He is paralysed in his lower spine ✓ • He is paralysed in his upper spine • The doctors want to increase the blood supply to his spine ✓ • The doctors want to decrease the blood supply to his spine • He has open heart surgery ✓	• One doctor wants to leave, the other doctors want her to go • One doctor wants to leave, the other doctors want her to stay ✓ • Two doctors want to leave • Derek regrets a decision he made ✓ • Dr Maggie Pierce is an alcoholic • Derek doesn't talk to his wife ✓
• 30/38 all correct answers (=79%)	• 10/38 all correct answers (=26%)

As shown in Table 2, more students understood what was happening medically to the characters (just under 80% of students got all the answers to the medical questions right) rather than what was going on in their personal lives and in the general plotline (just over a quarter of them got all these answers correct). The fact that this exercise took place in a medical English classroom perhaps skewed their answers as they concentrated more on the medical aspect of the video clip and paid less attention to the plot.

It can be argued that a medical student will naturally be more interested in the medicine being portrayed as opposed to the dramas of a fictional character's personal life. If this theory is correct, then it proves that a clip can be taken from a series, and its professional content be put to good use, without the students necessarily knowing the TV series and the characters involved.

A recurring question in both questionnaires and in the content and language test sought to evaluate the students' perception of hygiene and medical accuracy in audio-visual FASP by having them rate these on a scale of 0 to 5 (0=least accurate, 5=most accurate).

Table 3. Students' perception of medical accuracy

Accuracy scale	Pre-FASP (59)		During FASP (38)		Post-FASP (38)	
	No.	%	No.	%	No.	%
0	5	8.5	3	8	12	31.5
1	10	17	15	39.5	17	44.5
2	13	22	9	23.5	7	18.5
3	18	30.5	5	13	2	5.5
4	13	22	6	16	0	0
5	0	0	0	0	0	0

The results in Table 3 show an evolution between before and after the FASP sequence, with students rating medical TV shows as less accurate after the sequence and class discussion (over three quarters of students rated accuracy 0 or 1) than before it (over half of students rated accuracy 3 to 5). These results clearly show that the students developed their critical analysis skills through completing this task and that, by the final questionnaire, they were aware that the level of accuracy in medical television programmes is quite limited. This statement is supported by the students' reactions to this type of teaching support. Indeed, when asked about their opinion on audio-visual FASP in the post-FASP questionnaire, many of the students claimed that they appreciated using television shows to develop their critical analysis and as a way of providing visual images of bad practice – some spoke of learning through watching others' mistakes. They also found it an interesting way to learn medical vocabulary. However, the choice of the television show created some disagreement, as certain students felt that because of the lack of realism in the series, *Grey's Anatomy* was not a good pedagogical choice due to the many mistakes in the sequence. However, it can be argued that the use of fiction when examining medicine creates a distance from the professional situation which allows the students to develop their critical and analytical thinking, a skill which is often underdeveloped in other, more practical medical classes. Using a real-life documentary would more than likely generate less debate, as it would show a correct procedure undertaken by qualified professionals, therefore leaving little room for students to criticise. It is also important to remember that as most EMP teachers are not medical professionals, they need to be credible in the classroom, and FASP allows both teachers and learners to discuss a specific professional domain without necessarily being experts on the subject. Despite some reticence from the learners due to the choice of television series, the results clearly show the evolution of the students' critical analysis skills.

5. Discussion

The pilot study highlighted three main needs as expressed by medical students and professors at Nice University Medical Faculty. These findings are in line with other needs analysis EMP studies carried out in medical institutions in diverse geographical settings. First, our study showed a general acceptance of the necessity for English in medical studies which has also been reported in other needs analysis studies of medical students (Chia, Johnson, Chia, & Olive, 1999; Fang, 1987). Given the quantity of medical journals published in English nowadays – 4,609 in 2007 (Baethge, 2008) – the students' recognition of the importance of English is coherent and necessary. In addition, the pilot study underlined the importance of oral skills for medical students. This result confirms other studies but with a small difference. Chia et al. (1999) find that reading is the main skill needed by medical students, and Javid (2011) finds listening to be the most important skill according to questionnaire data, whereas reading is the first need expressed in other data collection tools in the same study. Both studies place speaking as the second most important skill for medical students. The variation in results obtained at Nice Medical Faculty in comparison to the two previously mentioned studies could be due to cultural differences – Chia et al.'s (1999) study took place in Taiwan and the study particularly mentions the students' timidity and lack of interest in communication skills. Alternatively, it could be the result of terminology choices; our study uses the term 'oral interaction' in opposition to Chia et al. (1999) and Javid (2011), where the more generic term 'speaking skills' is used. Further research into what specific oral interaction skills medical students need will have to be undertaken in order to understand this result more fully. Finally, students expressed a need for specific medical terminology, an opinion which was not at all shared by practising doctors who instead highlighted the need for cultural understanding. Indeed, in other EMP needs analysis literature the importance of cultural interaction is noted (Alqurashi, 2016), as well as the potential difficulty of certain contact situations for non-anglophone doctors (Allwright & Allwright, 1977; Ferguson, 2013), which include participating in informal discussion, and entertaining/ being entertained. When specific medical terminology is mentioned, it is as a secondary objective, with teaching from a medical and health care perspective being the priority (Antic, 2007). A needs analysis of higher level medical students is necessary to see the evolution of this desire for medical terminology and to measure at what stage the need for cultural understanding is expressed.

Through the use of audio-visual FASP, Study 2 allowed medical students to assimilate specific Anglo-Saxon medical culture practices whilst simultaneously developing their critical analysis skills and testing their medical vocabulary. It's precisely because audio-visual FASP gives access to professional environments in the target language and culture (otherwise inaccessible) that "professional behavioural culture" (Chapon, 2017, p. 48) can be witnessed. The didactic opportunity that audio-visual FASP represents, through the combination of specific professional culture skills and authentic discourse belonging to a particular professional domain, has already been highlighted by Chapon (2015) in the field of ELP. However, our study diverged from Chapon's research as we chose to use audio-visual FASP that did not faithfully reproduce specialist knowledge in order to check the students' comprehension of the tenets of good medical practice. To the best of our knowledge, no published research has investigated the efficacy of this approach in ESP settings so far. Still, in our study, the evolution of the students' perceptions of medical accuracy in the selected medical TV show points to the efficacy of this approach.

6. Conclusion

The first study identified the need for oral communication and cultural references in EMP, taking into consideration the opinions of professionals in the domain. Students assessed their needs as more vocabulary-based, although this is perhaps due to their professional inexperience and these demands will surely change as their practical knowledge increases. Indeed, the results show an evolution between L2 students (51% identified medical vocabulary as an important need) and L3 students (only 24.5% considered medical vocabulary to be important), and we can think that this need would further decrease with older groups. Inversely, the students did not particularly enjoy the cultural aspect of their medical English lessons (the second least favourite aspect of their course for L3 students), despite the medical professors' emphasis of this point. To reconcile students' perceived needs with the medical professors' perceived needs, FASP was used in the second study to raise students' awareness of the importance of good practice in English-speaking medical culture and to process a certain amount of medical language[14]. However, one of the limitations of this study was that the language learned during the FASP sequence was not evaluated at the end of the sequence to test what the students had retained from a purely linguistic point of view. During this sequence, the students' perception of the accuracy of *Grey's Anatomy* evolved significantly, which suggests that they became more critical in their analytical viewing of this medical TV series. In addition, audio-visual FASP had the advantage of giving clear examples of bad practice, enabling the students to visualise medical errors more easily in a dynamic and interesting way, although the choice of reverse pedagogy – showing what not to do in order to teach what should be done – was a risky one, as the students' reactions show. This second study proves how useful FASP can be in EMP as a more attractive and multi-disciplinary resource to learners, especially as regards cultural and professional awareness raising. Further research is needed, however, to test (1) the usefulness of audio-visual FASP in ESP in general (i.e. in other specialist domains) and (2) its impact on students' linguistic competence development.

References

Allwright, J., & Allwright, R. (1977). An approach to the teaching of medical English. In S. Holden (Ed.), *English for specific purposes* (pp. 58-62). Modern English Publications.

---

14. Babinski method, thoracic stent graft, aneurysm, upper, lower spine, etc.

Alqurashi, F. (2016). English for medical purposes for Saudi medical and health professionals. *Advances in Language and Literary Studies, 7*(6), 243-252.

Antic, Z. (2007). Forward in teaching English for medical purposes. *Facta Universitatis – Medicine & Biology, 14*(3), 141-147.

Baethge, C. (2008). The languages of medicine. *Deutsches Arzteblatt International, 105*(3), 37-40.

Belcher, D. (2004). Trends in teaching English for specific purposes. *Annual Review of Applied Linguistics, 24*, 165-186. https://doi.org/10.1017/S026719050400008X

Carnet, A. (2015). *Vers une didactique de l'anglais médical à visée professionnelle*. Journée d'étude DidASP, ESPE de Paris, 10 April.

Carnet, D., & Charpy, J.-P. (2017). Discours de professionnels et discours pour professionnels : le travail collaboratif au service de l'enseignement de l'anglais médical. *ASp, 71*, 47-68. https://doi.org/10.4000/asp.4952

Chapon, S. (2015). *Fiction à substrat professionnel télévisuel comme voie d'accès à l'enseignement/apprentissage de l'anglais juridique.* Unpublished thesis. Université Grenoble Alpes.

Chapon, S. (2017). Didactique de l'anglais juridique : de l'utilité des fictions judiciaires. *Les Langues Modernes, 3*, 47-52.

Charpy, J.-P. (2004). Milieux professionnels et FASP médicale : de l'autre côté du miroir. *ASp, 45-46*, 61-79. https://doi.org/10.4000/asp.866

Charpy, J.-P. (2005). La FASP médicale et ses marges : textes de références, prototextes et textes péripheriques. *ASp, 47-48*, 83-101. https://doi.org/10.4000/asp.795

Charpy, J.-P. (2010). FASP médicale et substrat professionnel : le miroir éclaté. *ASp, 57*, 61-79. https://doi.org/10.4000/asp.955

Charpy, J.-P. (2011). La FASP médicale comme outil pédagogique : authenticité des textes ou altération de l'authenticité ? *Les Cahiers de l'Apliut, 30*(2), 65-81. https://doi.org/10.4000/apliut.822

Chia, H.-U., Johnson, R., Chia, H.-L., & Olive, F. (1999). English for college students in Taiwan: a study of perceptions of English needs in a medical context. *English for Specific Purposes, 18*(2), 107-119. https://doi.org/10.1016/S0889-4906(97)00052-5

Czarny, M. J., Faden, R. R., Nolan, M. T., Bodensiek, E., & Sugarman, J. (2008). Medical and nursing students' television viewing habits: potential implications for bioethics. *The American Journal of Bioethics, 8*(12), 1-8. https://doi.org/10.1080/15265160802559153

Divasson, L., & León, I. (2002). Medical English and Spanish cognates: identification and classification. *ASp, 35-36*, 73-87. https://doi.org/10.4000/asp.1607

Ellis, R. (1994). *The study of second language acquisition*. Oxford University Press.
Ellis, R. (2003). *Task-based language learning and teaching*. Oxford University Press.
Ellis, R. (2013). Task-based language teaching: responding to the critics. *TESOL, 8*, 1-27.
English, K. (1998). Understanding science: when metaphors become terms. *Asp, 19-22*, 151-163. https://doi.org/10.4000/asp.2800
Fang, F. (1987). An evaluation of the English language curriculum for medical students. *Papers from the fourth conference on English teaching and learning in the Republic of China, 290*.
Faure, P. (2003). Formation des enseignants en langues de spécialité : exemple pour l'anglais médical. *Les Cahiers de l'Apliut, 22*(2), 9-27. https://doi.org/10.4000/apliut.3700
Ferguson, G. (2013). English for medical purposes. In B. Paltridge & S. Starfiled (Eds), *The handbook of English for specific purposes* (pp. 243-262). Wiley-Blackwell.
Foster, P. (1999). Task-based learning and pedagogy. *ELT Journal, 53*(1), 69-70. https://doi.org/10.1093/elt/53.1.69
Genty, S. (2010). La validation du substrat professionnel dans La proie de Michael Crichton (Prey, US, 2002). *ILCEA, 12*, 1-13.
Gledhill, C., & Kübler, N. (2016). What can linguistic approaches bring to English for specific purposes? *ASp, 69*, 65-95. https://doi.org/10.4000/asp.4804
Hoekje, B. (2012). Teaching English for medical and health professionals. In C. Chapelle (Ed.), *The encyclopedia of applied linguistics*. Blackwell. https://doi.org/10.1002/9781405198431.wbeal1154
Hutchinson, T., & Waters, A. (1987). *English for specific purposes*. Cambridge University Press. https://doi.org/10.1017/CBO9780511733031
Isani, S. (2004a). Popular films as didactic supports in ESP teaching – selection criteria and ethical considerations". In M. Petit (Ed.), *Aspects de la fiction à substrat professionnel. Collection travaux EA 2025* (pp. 121-132). Université Bordeaux 2.
Isani, S. (2004b). The FASP and the genres within the genre. In M. Petit (Ed.), *Aspects de la fiction à substrat professionnel. Collection travaux EA 2025* (pp. 25-36). Université Bordeaux 2.
Isani, S. (2006). Revisiting cinematic FASP and English for legal purposes in self-learning environment. *Cinéma et Langue de spécialité, Les Cahiers de l'APLIUT, 25*(1), 26-38. https://doi.org/10.4000/apliut.2575
Isani, S. (2011). Developing professional cultural competence through the multi-layered cultural substrata of FASP: English for legal purposes and M.R. Hall's The Coroner. *Cahiers de l'APLIUT, 30*(2), 29-45. https://doi.org/10.4000/apliut.1497

Javid, C. Z. (2011). EMP needs of medical undergraduates in a Saudi context. *Kashmir Journal of Language Research, 14*(1), 89-110.

Kramsch, C. (2000). Second language acquisition, applied linguistics, and the teaching of foreign languages. *Modern Language Journal,* 84(3), 311-326. https://doi.org/10.1111/0026-7902.00071

Maher, J. (1986). English for medical purposes. *Language Teaching, 19*(2), 112-145. https://doi.org/10.1017/S0261444800012003

Mourlhon-Dallies, F. (2008). *Enseigner une langue à des fins professionnelles.* Didier.

Pavel, E. (2014). Teaching English for medical purposes. *Bulletin of the Transilvania, 56*(2), 39-46.

Petit, M. (1999). La fiction à substrat professionnel : une autre voie d'accès à l'anglais de spécialité. *ASp, 23/26,* 57-81. https://doi.org/10.4000/asp.2325

Sarré, C., & Whyte, S. (2016). Research in ESP teaching and learning in French higher education: developing the construct of ESP didactics. *ASp, 69,* 139-164. https://doi.org/10.4000/asp.4834

Villez, B. (2004). Vers une didactique télévisuelle : Ally McBeal, la TASP et l'anglais de spécialité. In M. Petit (Ed.), *Aspects de la fiction à substrat professionnel. Collection travaux EA 2025* (pp. 103-111). Université Bordeaux 2.

Whyte, S., & Sarré, C. (2016). From 'war stories and romances' to research agenda: towards a model of ESP didactics. *ESSE, Galway, Eire, 22-26 August 2016.* https://www.slideshare.net/cherryenglish/war-stories-and-romances-whyte-sarr-esse-2016

10. The effect of form-focussed pre-task activities on accuracy in L2 production in an ESP course in French higher education

Rebecca Starkey-Perret[1], Sophie Belan[2], Thi Phuong Lê Ngo[3], and Guillaume Rialland[4]

Abstract

This chapter presents and discusses the results of a large-scale pilot study carried out in the context of a task-based, blended-learning Business English programme in the Foreign Languages and International Trade department of a French University[5]. It seeks to explore the effects of pre-task planned Focus on Form (FonF) on accuracy in students' written production. Using an action-research framework, the study consisted in introducing FonF pre-task activities in the programme and in analysing written productions of students with a B1 level. The researchers compared the results of the students who completed the form-focussed pre-tasks and those who did not complete these activities. The results show no significant differences in the productions of the control and experimental groups, leading the researchers to question the pertinence of pre-task FonF for B1 learners rather than post-task FonF, which can better cater for individual needs.

---

1. Université de Nantes, Nantes, France; rebecca.starkey1@univ-nantes.fr

2. Université de Nantes, Nantes, France; sophie.belan@univ-nantes.fr

3. Université de Nantes, Nantes, France; thi-phuong-le.ngo@etu.univ-nantes.fr

4. Université de Nantes, Nantes, France; guillaume.rialland1@univ-nantes.fr

5. In French higher education, Foreign Languages and International Trade departments and programmes are referred to as "Langues Étrangères Appliquées" or "LEA".

How to cite this chapter: Starkey-Perret, R., Belan, S., Lê Ngo, T. P., & Rialland, G. (2017). The effect of form-focussed pre-task activities on accuracy in L2 production in an ESP course in French higher education. In C. Sarré & S. Whyte (Eds), *New developments in ESP teaching and learning research* (pp. 171-195). Research-publishing.net. https://doi.org/10.14705/rpnet.2017.cssw2017.751

Chapter 10

Keywords: accuracy, computer-mediated learning, ESP, planned FonF, second language acquisition, SLA, task-based learning and teaching, TBLT.

1. Introduction

Perusal of practical guides for teachers (e.g. Ur, 2004), research papers in Second Language Learning (SLA) (e.g. Anderson, 2016; Long, 2014; Willis & Willis, 2009), as well as discussion with language teachers and future language teachers (Belan & Buck, 2012) quickly show that there is still ongoing debate in the world of language teaching on whether to propel learners into communicative situations and let them infer form from their communicative experiences (communicative approach) or rather to present preselected structures to be applied in controlled practice before taking the plunge into actual language use, or production – Presentation, Practice, Production (PPP) –, or perhaps to try to find a middle ground between FonF and communicative language use – Task-Based Language Teaching (TBLT).

Teachers' pedagogical choices tend to be based on their beliefs about how languages should be learned and taught (McAllister, Narcy-Combes, & Starkey-Perret, 2012). Some may repeat the modeling of their secondary school years, or rely on the official instructions provided (Narcy-Combes, 2005). It is recommended, however, to take a theory-grounded approach to teaching in order to update one's pedagogical choices according to the latest findings of research in SLA and in other fields that contribute to a more robust understanding of L2 learning and acquisition in institutional settings, such as psychology (educational, social, cognitive, etc.), sociology (Narcy-Combes, 2005), and, as far as English for Specific Purposes (ESP) contexts are concerned, language for specific purposes and content-based instruction, or Content and Language Integrated Learning (CLIL).

This chapter presents an action-research project carried out in the context of a Business English course in a French university. A group of teachers, researchers, and two postgraduate students have come together in an attempt to find common ground between theory-grounded practice, leading to a preference for TBLT, and in-service teacher representations clearly showing a preference for PPP and perceiving language teaching as explicit instruction of morpho-syntactic structures.

The results of the present study may be of particular interest to ESP practitioners and researchers in other contexts, as there seems to be quite a natural link between TBLT and ESP (Whyte, 2013). The TBLT framework is now commonly used in ESP courses as it allows to meet learners' specific needs (Dudley-Evans & St John, 1998) by introducing real-world tasks (Ellis, 2003) whose completion implies managing both subject-specific content and language to produce meaning (Llinares & Dalton-Puffer, 2015; Ortega, 2015; Whyte, 2016). This approach has been shown to help enhance language development in terms of fluency (McAllister & Belan, 2014). A study of the development of accuracy is also particularly relevant in ESP as learners' needs include acquiring professional competence in English, which is partly judged on language accuracy.

2. Description of the programme

The blended learning task-based programme was implemented in 2009, after two years of team cooperation (creation of materials, training, etc.) as an attempt to find solutions to the issues faced by the teaching team in their first year Business English classes: overcrowded groups (45 to 60 students) leading to limited individual feedback, lack of motivation and involvement, and high dropout rates (around 45%[6]). The necessity for individualisation became apparent, not only through the lens of SLA literature (Bygate, 2009; Robinson, 2002), but also through empirical data. A task-based written test was carried out at the beginning of the programme to assess students' individual levels according to the Common European Framework of Reference for languages (CEFR) (Council of Europe, 2001). The results showed that the English proficiency levels of first year

---

6. This figure reflects a common problem in language courses in French higher education. In 2012, 37.6% of language students left university after their first year (MESR, 2013, p. 2).

students at the University of Nantes are extremely heterogeneous, ranging from A1 to C2 (Buck & McAllister, 2011).

Taking a socio-constructivist and cognitivist/connectionist approach, the team designed a programme, to begin in the second semester of the first year, which combines classroom sessions with collaborative group work using a Moodle learning platform. In groups of three or four, students complete six to eight business-oriented, collaborative real-world tasks (R. Ellis, 2003) leading to oral and written productions. Corrective feedback is given in the form of advice and suggestions in order to help students 'notice' the gap (Schmidt & Frota, 1986) between their productions and the expected target language norms and pragmatic objectives. In the post-task phase, they are encouraged to complete form-focussed micro-tasks (Bertin, Gravé, & Narcy-Combes, 2010; Demaizière & Narcy-Combes, 2005) in a Virtual Resource Center (VRC) to focus their attention on their individual difficulties (Bygate, 2009; Robinson, 2002).

Concretely, the classroom sessions are organised into a one hour plenary (in theory 45 students) and one hour sessions in smaller groups of 15 students. The two hours that each student spends with the teacher are supplemented by two to four hours of computer-mediated work on the tasks outside of the classroom via three distinct spaces on a Moodle platform:

- 'The course space', which includes the scenarios and instructions for the preparatory work to carry out in order to complete each task, as well as additional resources and a course calendar.

- 'The class space', where students are enrolled manually by their teacher and which is organised as he/she wants. In this space, students submit the written task-productions for evaluation and feedback is given by the teachers. It includes a class forum, links to resources, reading materials, and homework corrections, etc.

- 'The VRC', which was built for form-focus troubleshooting. It contains contextualised practice exercises and explanations of specific forms as well as an explicit focus on specialised business vocabulary and pronunciation.

3. Theoretical underpinnings of the programme

The researchers involved in implementing the new programme adhered to the theoretical underpinnings of TBLT from a psychological perspective (socio-constructivist, cognitivist, and connectionist models of language acquisition) (Randall, 2007). A clear definition of how these theories of language acquisition relate to the programme will shed light on how student and teacher representations of the role of explicit grammar teaching have led them to experience cognitive dissonance with the programme and how the researchers have proposed to deal with the conflicting views on language and its acquisition in this specific context.

3.1. Connectionism, socio-constructivism, and cognitivism

In the connectionist framework, the object of language has been defined as a collection of patterns found in contextualised use of the target language by proficient users. According to this outlook, the rules of language are not prescriptive but usage-based, and therefore changes occur when speakers use language to communicate with each other, depending on the context (Lindquist, 2009). New models of how language is received, stored in memory, and then retrieved for use have emerged alongside the change in perception of the object of language itself. Connectionist accounts of language reception, storage, and retrieval can quite simply be put as follows: individuals notice frequent patterns in the input to which they are exposed and they store these data as prefabricated sequences in the brain which are then retrieved as chunks for use during communication (N. Ellis, 2003). This has an impact on pedagogy and on the way that form is dealt with in the institutional learning environment. Within the connectionist framework, learners of English wishing to communicate within specific contexts, instead of learning rules and then trying to apply them to communicative contexts, may study authentic language data to notice recurrent patterns of language use within targeted contexts in order to either infer rules from them, or to memorise them as formulaic chunks (Wray, 2002). This way of dealing with language structures is the one chosen by TBLT, and goes hand in hand with socio-constructivist and cognitivist paradigms of language learning.

In the socio-constructivist view, the role of interaction in the acquisition of L2 is primordial (Bruner, 2000; Vygotsky, 1978). This view is concerned with the interpsychological processes involved in learning and focusses on co-construction of knowledge. Cognitivism, on the other hand, is focussed on the intrapsychological processes which are triggered by the interaction. The central notions retained for the present context are temporarily directing learners' attention to form in context (Long, 1997) and individual practice (Robinson, 2002). The stance taken here is that in the institutional context, in which input is limited, it is possible to optimise input processing by helping the learners to direct their attention to salient features of the input, such as recurrent patterns/structures. This enables them to create hypotheses about the L2. In our programme, both authentic data in the pre-task phase (press articles on business issues, for example) and activities from the coursebook, *Market Leader intermediate*, are used. During these activities, learners are led to identify patterns and to practice the identified structures.

Attention will also be essential during output, whose importance for language learning has been highlighted by Swain (1985): the learner must pay attention to form in order to be understood by his or her interlocutor and is able to test and modify his or her hypotheses during communication until comprehension (or meaning) is reached. This process, known as negotiation of meaning (Swain, 2000), illustrates the essential link between form and meaning. The tasks used in our programme offer opportunities for negotiation of meaning during interaction as students must make decisions and solve problems[7].

After production, when the learner receives feedback, his/her attention will be focussed on noticing the gap between what he or she is able to produce in L2 and what he or she wanted to express; and/or the gap between what was said, and

---

7. See supplementary material parts 1-3 at https://research-publishing.box.com/s/uk3fc6fmax09odsk45z6h4s9wemzxbxz

the expected, most frequently found structures or collocations. According to the cognitivist framework, noticing must be supplemented with practice in order to make the language associations become more automatic (Ur, 2004). Because of individual differences in cognitive resources and capacity, this focussed practice should be done individually (Robinson, 2002), in the form of post-task form-focussed activities (Skehan, 1996; Skehan & Foster, 1997). Hence, within our programme, post-task form-focussed practice activities are proposed on the VRC, based on individual feedback.

Although the blended TBLT programme implemented is firmly rooted in SLA theory, one cannot overlook the weight of student and teacher representations and affect in the success or failure of a given pedagogical programme. If a student perceives the programme as being useful for language learning and for his/her professional objectives, he or she will more likely invest in the programme than if he or she perceives it as being inefficient (Whyte, 2013; Wigfield & Eccles, 2000). In the same vein, if a teacher does not identify with the underlying principles behind a programme, or does not believe in the students' abilities, their engagement could be hindered (McAllister et al., 2012) and/or their lack of enthusiasm could be contagious (Hatfield & Cacioppo, 1994).

3.2. Student and teacher representations

Previous studies on students' and teachers' representations (McAllister & Narcy-Combes, 2015; McAllister et al., 2012; Narcy-Combes & McAllister, 2011; Starkey-Perret, McAllister, & Narcy-Combes, 2012) on the development of accuracy, fluency, and complexity of written production between the beginning and the end of the programme (McAllister, 2013; McAllister & Belan, 2014) and on the students' use of the virtual resource center (McAllister, 2013; Starkey-Perret, McAllister, Belan, & Ngo, 2015) showed that although the programme is generally appreciated for the opportunities it generates for small-group interaction (McAllister et al., 2012; Starkey-Perret et al., 2012), students and teachers tend to prefer a PPP approach, claiming that TBLT does not leave sufficient room for FonF (Belan & Buck, 2012; McAllister, 2013). Questionnaire studies carried out with the students showed that the way 'grammar' is dealt with is the least satisfactory element of the programme and that there simply is not enough FonF (Belan & Buck, 2012). It seems that many of the in-service teachers and the students in the present study take a symbolic view of language in which it is perceived as a set of rules to be memorised and then applied to production. Informal discussions with teachers during meetings in 2015 and 2016 reveal general dissatisfaction among the in-service teachers. Recurrent comments include "it doesn't work", "their English is getting worse", "they don't do the work anyway" and "in Spanish they actually work, because they have grammar classes".

Studies also showed that the VRC is underexploited by the students, which implies that they do not do the individual post-task form-focussed activities, which makes it difficult to assess the VRC's effects on language development (McAllister, 2013; Starkey-Perret et al., 2015). A questionnaire study showed that just under half of the students (48%) visited the VRC, 66% of whom used it between one and three times over the course of the semester for a duration of under 30 minutes each time (Starkey-Perret et al., 2015). However, it was noted in the same study that the students who used the VRC found it useful for learning and that the more they used it, the more they found it useful.

Another questionnaire concerning the VRC was distributed to the 12 teachers involved in the programme but only four responded, two of whom were also researchers involved in setting up the programme and who claimed to encourage their students to visit the VRC during the post-task phase. The other two respondents declared that they did not advise the students to use the VRC for individual troubleshooting and practice because "they would not do it anyway". Here, the weight of teachers' representations of students in their pedagogical choices, and the negative Pygmalion effect (Rosenthral & Jacobsen, 1968), or self-fulfilling prophecy, this can generate, is clearly demonstrated. Further comments concerning the computer-mediated aspect of the programme include "they don't have enough hours of real class time" and "a computer doesn't replace a teacher". For the majority of the teachers involved, the only solution was to abandon the programme entirely in order to reinstitute face-to-face 'grammar' classes for first year students.

These results show that the application of the blended learning task-based programme for this Business English course has not been entirely successful, even though positive results have been found. The resistance with which the programme was met led the researchers in the team, clearly involved in an action-research perspective, to go back to the drawing board and find ways to modify the programme so as to make its application more palatable to all of the users involved, all the while trying to remain coherent with current research in SLA, one of the greatest challenges of large-scale classroom-based research.

3.3. Exploring the potential of interface between declarative knowledge and automatised use of L2

The main area that necessitated our attention to SLA theory was the place of FonF in meaning-focussed programmes and how this place relates to declarative and procedural memory and knowledge, which in turn affects learners' abilities to use language in real-life communication. A lack of consensus in the field led us to choose the perspective that we believed would be the most coherent with a TBLT course in an ESP context, all the while making the necessary concessions to facilitate acceptance of our programme by the teachers and the students who have a preference for PPP.

We decided to follow the weak interface position presented by Ellis (2002) as a compromise between Krashen's (1981) non-interface position and a strong interface position. In the non-interface position, it was noted that learning explicit grammar rules enables learners to improve their explicit knowledge of English grammar and leads them to better perform on activities requiring explanation of language rules or demonstration of explicit knowledge of grammar. However, this type of declarative knowledge does not enable learners to better perform on activities requiring the use of language in real-life communicative contexts (Maraco & Masterman, 2006). In other terms, a learner can study grammar rules for years on end without ever being able to produce real-world language with a high-degree of accuracy, fluidity, or complexity. In the strong interface position, explicit knowledge becomes automatised through practice, just like any other skill, such as driving a car.

If this is the case, then why is it that after eight years of English study many learners, although good at reciting grammar rules, cannot seem to use them when communicating under the constraints of online processing of language during actual language use? If it is simply that they were not given ample opportunity to practise, then a strong case has been made for PPP, as long as the final P is given full attention: "the importance of the production phase, which is often shortened or omitted in practice (e.g. Choi & Andon, 2014; Sato, 2010), should be emphasised" (Anderson, 2016, p. 20).

In Ellis's (2002) weak interface position, explicit and implicit knowledge do interact in some way in long term memory but it is unknown to what extent. Students who are involved in meaning-focussed language programmes which incorporate form-focussed activities as moments of 'time-out' (Belan & Buck, 2012) when encountering a problem during task-preparation attain higher degrees of accuracy. This position seemed the most coherent to our programme as it is the basis for the concepts of noticing and attention central to TBLT. Furthermore, it concorded with research carried out by psychologists specialised in memory such as Baddeley, Eysenk, and Anderson (2009) who established that explicit knowledge cannot become procedural, but that explicit knowledge can facilitate the acquisition of procedural knowledge and that the two systems are built up at the same time during real-life language use.

3.4. Bridging the gap: finding the right point of convergence

In order to find applicable solutions, we decided to seek out where the points of view of the students, the teachers, and the researchers converged. The point of convergence seems to be the following: focussing only on content is insufficient to attain high-degrees of accuracy during production, and learning can be accelerated and optimised in adult learners if they periodically direct their attention to form (Long, 1997). It was previously thought that, in our programme, the periodic FonF was being carried out via post-task form focus on the VRC, as originally planned when designing the course, but studies on the use of the VRC indicated that this was not the case.

The compromise that was found after determining that we accept, for the current study, a weak interface position, was to introduce contextualised form-focussed pre-task activities via the use of the VRC. Giving the place of FonF to the pre-task phase definitely feels more PPP than TBLT. However, finding a common ground with student and teacher representations seemed essential.

> "As such, PPP may or may not be an accurate representation of how languages are learnt on an individual level, but it reflects well how many of us expect to be taught a new skill on a social level (Borg, 1998; Burgess & Etherington, 2002; Widdowson, 1990)" (Anderson, 2016, p. 16).

However, the way in which to go about this needed to be carefully construed. According to Ellis and Shintani (2014),

> "explicit grammar instruction has a place in language teaching but not based on a grammatical syllabus. Instead, it should draw on a checklist of problematic structures and observational evidence of their partial acquisition" (p. 112).

This led us to study student productions in order to pinpoint recurrent problems.

4. Research questions

Following our action-research orientation, it was deemed necessary to adapt the programme to the participants' desire for form-focussed instruction by adding pre-task, computer-mediated form-focussed activities. In this study, we sought to assess the effectiveness of these activities on accuracy in students' written production. Hence, the following research questions were formulated:

- What effects of completing pre-task, form-focussed, computer-mediated activities can be observed on **frequency** of use of targeted forms?

- What effects of completing pre-task, form-focussed, computer-mediated activities can be observed on **accuracy** of use of targeted forms?

5. Planned pre-task FonF activities: procedure

For the current study, three of the eight macro-tasks were selected for integration of pre-task FonF. In order to determine which structures would be targeted upstream, two procedures were used simultaneously. On the one hand, the forms students need to be able to use in order to complete the tasks successfully were identified, hence maintaining the link between form and meaning. On the other hand, we studied a small sample (n=20) of student productions on the same tasks from the previous year to identify recurring problems. Once the problematic forms were identified, preparatory micro-tasks were created via the software LearningApps and Quizlet, and integrated in the platform. Additionally, pre-existing micro-tasks on the VRC dealing with the selected forms were identified. The form-focussed micro-tasks led learners to focus on the pre-selected forms to identify (notice) them in short authentic documents (input enhancement), categorise them, and use them in controlled exercises and games that are corrected automatically. Links to contextualised explanations of the forms were incorporated within the activities for the learners who prefer to be able to 'read the rule'. The software programmes chosen were those that seemed the most ergonomic: the automatic feedback is easy to access, and links to further explanations and information were easy to integrate for the developers and easy to locate for the students. Furthermore, they offer opportunities to set up the micro-tasks in game form, which was deemed more motivating for the students. However, data has not been collected concerning the students' perception of the software, leading to a bias in the current study.

Altogether, 18 activities of pre-task form focus were included (Table 1). Macro-task instructions were rewritten to include links to the form-focussed pre-task activities in the VRC. A bias may have been introduced by the fact that the number of micro-tasks for each task varies. It was hoped that by the time students reached the third task, they would already be familiar with the procedure and would have seen the benefit of pre-task FonF. It should be noted here that students had been working with the platform for a semester before participating in the current study.

Table 1. Description of the planned FonF activities

Theme	Macro-task	Identified forms	Types of activities
International Markets	Carrying out market research for a chosen country to analyse the opportunities and threats associated with doing business there, then writing a report to present the results[8]	• connectives • adverbs • adjectives	• identification of forms while reading authentic marketing documents • categorising forms • classifying forms according to their function in a sentence.
Ethics	Writing a personal narrative[9]	• past tenses • prepositions of time and place	• identification of forms • classifying forms according to their function • cloze tests
Competition	Carrying out analysis of the French market and the potential competition for Costa Coffee, then summing up the findings and making recommendations in a formal written report[10]	• definite/indefinite articles • demonstratives • possessives • quantifiers • phrasal verbs • comparatives • superlatives • relative pronouns • modals • conditionals	• identification of forms • classifying forms according to their function • categorising forms • linking phrasal verbs to definitions • cloze tests

8. See supplementary material part 1 for the scenario and task instructions: https://research-publishing.box.com/s/uk3fc6fmax09odsk45z6h4s9wemzxbxz

9. See supplementary material part 2 for the scenario and task instructions: https://research-publishing.box.com/s/uk3fc6fmax09odsk45z6h4s9wemzxbxz

10. See supplementary material part 3 for the scenario and task instructions: https://research-publishing.box.com/s/uk3fc6fmax09odsk45z6h4s9wemzxbxz

6. Procedure

6.1. Data collection

The students' online activity was monitored over the course of a 12-week semester. The researchers could see who engaged in the pre-tasks and which activities were completed. Then, three written productions, corresponding to the three macro-tasks of the study, per student were collected and analysed to identify if the targeted forms were used, if the targeted forms used were consistent with L2 norms, and if the completion of the online pre-task activities had any effect. Additionally, the entire group of first year students (n=564) took a pre-intervention test to control for proficiency. The test showed that 337 students (60%) were at B1 level. As previous research showed that the blended TBLT programme was more effective for B1 learners than for A2 or B2 learners (Buck & McAllister, 2011), the researchers decided to focus only on this sub-group.

6.2. Sampling: control group and experimental group

One of the biggest challenges in classroom-based research is the establishment of control groups and experimental groups in order to increase the interpretative value of the results. In our study, students were assigned to control and experimental groups by self-selection: all were offered pre-task activities, but only some chose to complete them. For each of the three tasks, some students chose not to complete all the preparatory activities, hence the differences that appear in the tables presented below. We acknowledge a bias in our research with the possibility that the students in the experimental group were more motivated and spent more time overall on tasks, but felt that it was not ethically acceptable to use random assignment to one condition or the other.

7. Results

Table 2 shows that the number of students who completed the micro-tasks varied over the course of the semester. The students participated the most in the micro-tasks related to the first task of the study. There was a severe drop in the numbers of those who completed the micro-tasks for the second task. A closer look at the data showed that for two of the teachers in the programme, none of the students completed the micro-tasks, which may indicate that for the second task, those teachers did not remind their students to complete the preparatory activities online. Concerning the micro-tasks for the third task, a clear drop in participation can be seen between the first activities proposed and the subsequent ones. This may indicate that students lost interest in the activities before completing them.

Table 2. Number of students who completed the online micro-tasks and the tasks

		Micro-tasks and task	Task only
T1 (192 productions)	Connectives	67	125
	Adjectives & adverbs	70	122
T2 (71 productions)	Simple past	18	53
	Prepositions of time & place	12	59
T3 (101 productions)	Determiners (definite & indefinite articles, quantifiers, possessives, demonstratives)	36	65
	Comparatives & superlatives	7	94
	Relative pronouns	8	93
	Modals	9	92

Based on the forms identified by the researchers for each task, the analysis of students' written productions consisted in determining the number of occurrences of each form and the number of these occurrences that conformed with the L2 norms. The following paragraphs focus on each task and on each group's performance in terms of number of occurrences and levels of conformity to L2 norms.

7.1. Task 1 – writing a report to present the results of market research

In Table 3, the data show that the group of students who completed the micro-tasks (A) in Task 1 produced slightly more occurrences of each targeted form than those who did not (B), however, the percentages of conformity for each form are very similar in both groups.

Table 3. Analysis of students' productions, task 1

	Connectives		Adjectives		Adverbs	
Group	A	B	A	B	A	B
n =	67	125	70	122	70	122
average nb of occurrences	37.4 SD=16.6	36.1 SD=14.7	58.4 SD=24.8	55.1 SD=24.3	20.7 SD=9.9	18.1 SD=7.8
average % of conformity	97.6% SD=3.2	97.5% SD=3.6	91.1% SD=5.9	91% SD=8.2	96.4% SD=5.4	96.7% SD=5.1

7.2. Task 2 – writing a personal narrative

This second task was designed so that students could use the simple past and prepositions of time and place. As shown in Table 4, the students who had completed the micro-tasks on the simple past produced slightly fewer occurrences of the targeted tense and their results in terms of conformity are very similar to those of the students who did not do the pre-task activities.

As for prepositions of time and place, the students who had completed the micro-tasks performed better and produced more occurrences than the students who did not. Their productions show a higher percentage of appropriate use for prepositions of time. However, it should be noted that the differences between groups are not statistically significant. A comparison of group means shows that the intergroup differences all fall within the standard deviation, which is sufficient to determine their lack of statistical significance.

Table 4. Analysis of students' productions, Task 2

	Simple Past		Prepositions of time		Prepositions of place	
Group	A	B	A	B	A	B
n =	18	53	12	59	12	59
average nb of occurrences	18 SD=6.8	22.4 SD=11.8	3.3 SD=1.8	2.5 SD=2	9 SD=3.9	7.6 SD=5.4
average % of conformity	89.3% SD=9	89.1% SD=16.4	90.1% SD=15.5	86.8% SD=26.3	84.5% SD=14.6	83.7% SD=20

7.3. Task 3 – summing up the findings of market research and making recommendations in a formal report

Task 3 was the most complex and the most demanding of the three tasks examined in this study. For this task, the researchers identified 11 forms to focus on (Table 5). Two of them – phrasal verbs and conditionals – were used by very few students in their productions (less than one occurrence on average) so they were not taken into account for this study as no significant data could be exploited. This may show that the forms the researchers deemed necessary for task completion could be avoided by the students, without impeding their capability to successfully complete the task.

Table 5. Analysis of students' productions, Task 3

	Definite article		Indefinite article		Quantifiers		Possessives		Demonstratives	
Group	A	B	A	B	A	B	A	B	A	B
n=	36	65	36	65	36	65	36	65	36	65
average nb of occurrences	41.9 SD = 13.3	39.1 SD = 20.6	19.6 SD = 11.3	14.8 SD = 7.8	2.3 SD = 2.1	2 SD = 2.1	3.2 SD = 1.9	4.9 SD = 4.3	3.7 SD = 2.3	3 SD = 1.7
average % of conformity	87.9% SD = 4.9	87.9 SD = 10.1	89.6% SD = 6.7	86.8% SD = 13	79.3% SD = 25.8	67% SD = 40.6	88.4% SD = 26.1	96.3% SD = 7.4	85.5% SD = 17.7	93.8% SD = 14.6

	Comparatives		Superlatives		Relative pronouns		Modals	
Group	A	B	A	B	A	B	A	B
n=	7	94	7	94	8	93	9	92
average nb of occurrences	2.3 SD=2.0	3.3 SD=2.5	2 SD=1.5	1.7 SD=1.6	8 SD=3.9	4.6 SD=3.5	18.7 SD=2.6	12.6 SD=5.9
average % of conformity	100% SD=0	76.8 SD=30.5	80% SD=16.3	88.8% SD=18.6	97.5% SD=6.6	88.2% SD=25.7	98.1% SD=2.8	97.6% SD=6.6

In the determiners category (Table 5), which comprises five different forms, articles were naturally the most frequently used. For possessives and demonstratives, the number of occurrences of each form was quite low (less than five occurrences per production on average), with more possessives used by the students who chose not to do the preparatory tasks and higher conformity in the use of both forms. The second most frequent form in this third task was modals. The difference between the two groups is quite clear as far as quantity is concerned, but the percentages of conformity of both groups are very close, with a difference of only 0.51%. However, as for the previous results, a comparison between means and standard deviations shows that these differences cannot be considered statistically significant.

8. Discussion

The study shows that it is indeed possible to create form-focussed pre-tasks, but it is difficult to anticipate their effects on student engagement and on language acquisition. Looking at the research questions (what effects on frequency and accuracy of these forms are observed?), we note that out of the 15 targeted forms presented in the results, 12 were found more frequently in the productions of students having done the pre-tasks. Caution must be used, however, when interpreting the results, as the difference in frequency of use of the targeted forms in the two groups is often very slight and none of the differences observed between groups is statistically significant. Frequency of use of a form is interesting from an acquisitional standpoint, as it highlights the necessary risk-taking in using new forms involved in the process of language development. However, this does not always lead to higher degrees of accuracy of these forms during production, indicating perhaps that their acquisition is not yet fully attained. The present study did not show that there was any significant gain in accuracy for the group of students who completed the pre-tasks. To get more precise and significant data about the efficiency of FonF pre-task activities for the development of accuracy, further studies could focus on fewer variables (one or two grammar points) over a longer period of time (for example an entire academic year).

The results of this study could be used to destabilise teacher representations of language learning and teaching and to reinforce the argument for individualised *post-task* FonF as originally planned in the programme: teachers were asked to give corrective feedback in the form of advice and suggestions to complete form-focussed micro-tasks in the VRC, but did not do it as this procedure did not correspond to their representations of language teaching and learning.

This leads us to reiterate the necessity of fostering opportunities for individualisation, especially in large-group settings, which is exactly what the hybrid language classroom enables. Offering pre-task form-focussed activities may not be useless. However, gains in accuracy were not shown systematically in the productions of the experimental group, and none of the gains were statistically significant. Pre-task form focus may be most beneficial if the students are not forced to do the activities but *choose* to do them because they realise that the task demands command of specific forms in order to express desired meanings. This is coherent with research in SLA and in TBLT showing that language learning is enhanced when the learner becomes aware of the gaps that exist between what they *would like* to communicate and what they *are able* to communicate. For each learner these gaps will be different, hence the benefit in offering up a wide variety of FonF activities that they can choose from depending on their individual needs at a given-time (pre- or post-task).

Additionally, we are able to partially analyse what users of platforms actually do with the resources compared to the way the developers imagined the resources would be used. Many previous studies have shown discrepancies between developers' envisioned use of the resources and actual use by the learners (see Andrianirina & Foucher, 2007; Docq & Daele, 2003; Fischer, 2012). We noted for example, that the students' engagement in the tasks decreased throughout the semester and remained rather low. This is a common problem in first year language courses in French universities, but there may be other reasons that are not currently measured, such as the types of pre-task activities proposed (game, gap-fill, etc.), the software used, the students' representations of whether or not they need to focus on said form, the students' representations of their ability to tackle said forms, and the involvement of each teacher.

Other factors may also need to be taken into account, like the students' study level and the characteristics of their study programme, Foreign Languages and

International Trade. These language programmes aim at training students in at least two foreign languages applied to various professional domains within the business and trade sphere (e.g. economics, marketing, management). Although it can be argued that Business English in Foreign Languages and International Trade is "one of the branches of ESP" (Narcy-Combes, 2008, p. 133), it is sometimes not considered a language for specific purposes course like those designed for 'specialists of other disciplines' (Van der Yeught, 2014), which are often more ESP-oriented given the students' profiles (e.g. students at Economics faculties) and the more specific professional domains they relate to.

This study enabled us to further validate previous studies showing that a majority of students upon entry to university do not have the institutionally targeted B2 level (Buck & McAllister, 2011; Frost & O'Donnell, 2015). In their first year, Foreign Languages and International Trade students may have similar linguistic and pragmatic needs as students specialised in other disciplines, but these needs could be different to some extent as students may lack subject knowledge (business-related issues) and lexis, both in L1 and L2. This probably results in a lack of motivation to explore a specialised domain and could explain their low level of engagement in the Business English programme. A similar study involving ESP learners at the same level or more advanced Foreign Languages and International Trade students (Master's students for example) could help to determine whether a higher level of specialisation has an impact on student engagement in pre-task form-focussed activities and in contextualised tasks in general.

Further research will focus on who actually uses the resources the most (A2, B1, B2, or C1 learners?), and when they use them (pre-task or post-task). Is it those who need them the most in order to succeed in their first year Business English courses (A2/B1 learners)? Or is it those who need them the least (B2/C1 learners)? Previous research has indicated that those who have higher levels are those who tend to use them the most as they are also those who tend to be the most autonomous in their learning (McAllister, 2013; Prince, 2009).

Further pedagogical development will focus on enriching the VRC so as to offer students an extensive database of contextualised activities on a wide variety of problematic structures, adapted to all levels of the CEFR. Beyond developing the resources, making the students more aware of their failings in performing the different tasks may help them understand the importance of having more autonomous attitudes toward language learning, and may encourage them to use the resources available to them.

9. Conclusion

This study shows that there were no statistically significant differences in accuracy of use of targeted forms between the students who had completed the form-focussed pre-tasks and those who had not. This reinforces recommendations of post-task FonF post-task based on each learner's individual needs. Greater focus should then be given to destabilising student and teacher representations of language learning and guiding students towards more autonomous behaviours related to language learning in order to foster their engagement in individualised post-task FonF.

Further research could also help determine the forms students at each level of the CEFR specifically need to focus on, thus helping the researchers and practitioners develop more relevant micro-tasks in the VRC, so as to cater for students' individual needs more efficiently, both in pre-task and post-task phases.

References

Anderson, J. (2016). Why practice makes perfect sense: the past, present and potential future of the PPP paradigm in language teacher education. *ELTED, 19*, 14-22.
Andrianirina, H., & Foucher, A.-L. (2007). Comment les usages réels d'un dispositif d'apprentissage de l'anglais à distance sont-ils perçus par les concepteurs-animateurs du dispositif? *Actes de la journée scientifique Rés@tice 2007, 13-14 décembre 2007, Rabat, Maroc*. http://www.youscribe.com/BookReader/Index/530973/?documentId=502096
Baddeley, A. D., Eysenk, M. W., & Anderson, M. C. (2009). *Memory*. Psychology Press.

Belan, S., & Buck, J. (2012). Attending to Form in a meaning-focused programme: the integration of grammar into a task-based blended learning programme. *Etudes en Didactique des Langues n°20 - Quelle grammaire en LANSAD? / What grammar for ESOL?*, Universisté de Toulouse - Le Mirail.

Bertin, J.-C., Gravé, P., & Narcy-Combes, J.-P. (2010). *Second language distance learning and teaching: theoretical perspectives and didactic ergonomics*. Hershey. https://doi.org/10.4018/978-1-61520-707-7

Borg, S. (1998). Talking about grammar in the foreign language classroom. *Language Awareness*, 7(4), 159-175. https://doi.org/10.1080/09658419808667107

Bruner, J. (2000). *Culture et modes de pensée : l'esprit humain dans ses œuvres*. Editions Retz.

Buck, J., & McAllister, J. (2011). Mise en place d'un dispositif d'apprentissage hybride à l'université. *Les Cahiers de l'APLIUT*, 30(1). https://doi.org/10.4000/apliut.571

Burgess, J., & Etherington, S. (2002). Focus on grammatical form: explicit or implicit? *System*, 30(4), 433-458. https://doi.org/10.1016/S0346-251X(02)00048-9

Bygate, M. (2009). Effects of task repetition on the structure and control of oral language. In K. Van den Branden, M. Bygate & J. M. Norris (Eds), *Task-based language teaching: a reader* (pp. 249-274). John Benjamins.

Choi, T., & Andon, N. (2014). Can a teacher certification scheme change ELT classroom practice? *ELT Journal*, 68(1), 12-21. https://doi.org/10.1093/elt/cct059

Council of Europe. (2001). *Common European framework of reference for languages. Learning, teaching, assessment*. Cambridge University Press.

Demaizière, F., & Narcy-Combes, J.-P. (2005). Méthodologie de la recherche didactique : nativisation, tâches et TIC. *ALSIC*, 8(1), 45-64. https://doi.org/10.4000/alsic.326

Docq, F., & Daele, A. (2003). De l'outil à l'instrument : des usages en émergence. In B. Charlier & D. Peraya (Eds), *Technologie et innovation en pédagogie : dispositifs innovants de formation pour l'enseignement supérieur* (pp. 113-128). De Boeck.

Dudley-Evans, T., & St John, M. (1998). *Developments in English for specific purposes: a multi-disciplinary approach*. Cambridge University Press.

Ellis, N. (2003). Constructions, chunking, and connectionism: the emergence of second language structure. In C. J. Doughty & M. H. Long (Eds), *The handbook of second language acquisition* (pp. 63-103). Blackwell. https://doi.org/10.1002/9780470756492.ch4

Ellis, R. (2002). The place of grammar instruction in the second/foreign language curriculum. In S. Fotos & E. Hinkel (Eds), *New perspectives on grammar teaching in second language classrooms* (pp. 17-34). Lawrence Erlbaum Associates, Inc.

Ellis, R. (2003). *Task-based language learning and teaching*. Oxford University Press.

Ellis, R., & Shintani, N. (2014). *Exploring language pedagogy through second language acquisition research*. Routledge.

Fischer, R. (2012). Diversity in learner usage patterns. In G. Stockwell (Ed.), *Computer-assisted language learning: diversity in research and practice* (pp. 14-32). Cambridge University Press. https://doi.org/10.1017/CBO9781139060981.002

Frost, S., & O'Donnell, J. (2015). Success: B2 or not B2 - that is the question. *Recherche et pratiques pédagogiques en langues de spécialité, 35*(2). https://doi.org/10.4000/apliut.5195

Hatfield, E., & Cacioppo, J. T. (1994). *Emotional contagion*. Cambridge University Press.

Krashen, S. (1981). *Second language acquisition and second language learning*. Pergmamon.

Lindquist, H. (2009). *Corpus linguistics and the description of English*. Edinburgh University Press.

Llinares, A., & Dalton-Puffer, C. (2015). The role of different tasks in CLIL students' use of evaluative language. *System, 54*, 69-79. https://doi.org/10.1016/j.system.2015.05.001

Long, M. (1997). *Focus on form in task-based language teaching*. http://www.mhhe.com/socscience/foreignlang/conf/first.htm

Long, M. (2014). *Second language acquisition and task-based language teaching*. Wiley.

Maraco, E., & Masterman, L. (2006). Does intensive explicit grammar instruction make all the difference? *Language Teaching Research, 10*(3), 297-327. https://doi.org/10.1191/1362168806lr197oa

McAllister, J. (2013). Évaluation d'un dispositif hybride d'apprentissage de l'anglais en milieu universitaire : potentialités et enjeux pour l'acquisition d'une L2. Doctoral thesis. Université de Nantes.

McAllister, J., & Belan, S. (2014). L'anglais de spécialité en LEA à la croisée des domaines : étude de l'acquisition du lexique spécialisé. *ASp, 66*, 41-59. https://doi.org/10.4000/asp.4564

McAllister, J., & Narcy-Combes, M.-F. (2015). Étude longitudinale d'un dispositif hybride d'apprentissage de l'anglais en milieu universitaire – le point de vue des étudiants. *Alsic, 18*(2). http://alsic.revues.org/2858

McAllister, J., Narcy-Combes, M.-F., & Starkey-Perret, R. (2012). Language teachers' perceptions of a task-based learning programme in a French University. In A. Shehadeh & C. A. Coombe (Eds), *Task-based language teaching in foreign language contexts: research and implementation* (pp. 313-342). John Benjamins. https://doi.org/10.1075/tblt.4.18mca

MESR. (2013). *Ministère de l'enseignement supérieur et de la recherche – Note d'information n°13.10: "Réussite et échec en premier cycle"*. Paris: DEPP-DVE. http://cache.media.enseignementsup-recherche.gouv.fr/file/2013/44/7/NI_MESR_13_10_283447.pdf

Narcy-Combes, M.-F. (2005). *Précis de didactique – devenir professeur de langue*. Ellipses

Narcy-Combes, M.-F. (2008). L'anglais de spécialité en LEA : entre proximité et distance, un nouvel équilibre à construire. *ASp, 53-54*, 129-140. https://doi.org/10.4000/asp.396

Narcy-Combes, M.-F., & McAllister, J. (2011). Evaluation of a blended language learning environment in a French university and its effects on second language acquisition. *ASp 59*, 115-138. https://doi.org/10.4000/asp.2250

Ortega, L. (2015). Researching CLIL and TBLT interfaces. *System, 54*, 103-109.

Prince, P. (2009). Un ménage à trois fragile : autonomie, motivation et apprentissage dans un centre de langues. *LIDIL, 40*, 71-88. http://lidil.revues.org/2925

Randall, M. (2007). *Memory, psychology and second language learning*. John Benjamins Publishing Company. https://doi.org/10.1075/lllt.19

Robinson, P. (2002). *Individual differences and instructed language learning*. John Benjamins Publishing Company. https://doi.org/10.1075/lllt.2

Rosenthral, R., & Jacobsen, L. (1968). Pygmalion in the classroom. *The Urban Review, 3*(1), 16-20. https://doi.org/10.1007/BF02322211

Sato, R. (2010). Reconsidering the effectiveness and suitability of PPP and TBLT in the Japanese EFL classroom. *JALT Journal, 32*(2), 189-200.

Schmidt, R., & Frota, S. N. (1986). Developing basic conversational ability in a second language: a case study of an adult learner of Portuguese. In R. Day (Ed.), *Talking to learn: conversation in second language acquisition* (pp. 237-326). New House.

Skehan, P. (1996). A framework for the implementation of task-based instruction. *Applied linguistics, 17*(1), 38-62. https://doi.org/10.1093/applin/17.1.38

Skehan, P., & Foster, P. (1997). Task type and task processing conditions as influences on foreign language performance. *Language Teaching Research, 1*(3), 185-211.

Starkey-Perret, R., McAllister, J., Belan, S., & Ngo, T. P. L. (2015). Assessing undergraduate student engagement in a virtual resource center. *Recherche et pratiques pédagogiques en langues de spécialité, 34*(2). https://doi.org/10.4000/apliut.5184

Starkey-Perret, R., McAllister, J., & Narcy-Combes, M.-F. (2012). Représentations des enseignants d'anglais et évaluation d'un dispositif hybride : image de soi, image de l'apprenant et appropriation du dispositif. *Cahiers de l'APLIUT, 31*(1), 74-96. https://doi.org/10.4000/apliut.2321

Swain, M. (1985). Communicative competence: some roles of comprehensible input and comprehensible output in its development. In S. M. Gass & C. Madden (Eds), *Input in second language acquisition* (pp. 235-253) Newbury House.

Swain , M. (2000). The output hypothesis and beyond; mediating acquisition through collaborative dialogue. In J. Lantolf (Ed.), *Sociocultural theory and second language learning* (pp. 97-114). Oxford University Press.

Ur, P. (2004). *A course in Language teaching*. Cambridge University Press.

Van der Yeught, M. (2014). Développer les langues de spécialité dans le secteur LANSAD – scénarios possible et parcours recommandés pour contribuer à la professionnalisation des formations. *Recherches et pratiques pédagogiques en langues de spécialité, 33*(1). https://doi.org/10.4000/apliut.4153

Vygotsky, L. S. (1978). *Mind in society: development of higher psychological processes*. University of Harvard Press.

Whyte, S. (2013). Teaching ESP: a task-based framework for French graduate courses. *ASp, 63*, 5-30. https://doi.org/10.4000/asp.3280

Whyte, S. (2016). Who are the specialists? Teaching and learning specialised language in French educational contexts. *Recherches et pratiques pédagogiques en langue de spécialité, 35*(3). https://doi.org/10.4000/apliut.5487

Widdowson, H. G. (1990). *Aspects of language teaching*. Oxford University Press.

Wigfield, A., & Eccles, J. S. (2000). Expectancy-value theory of achievement motivation. *Contemporary Educational Psychology, 25,* 68-81. https://doi.org/10.1006/ceps.1999.1015

Willis, D., & Willis, J. (2009). Task-based language teaching: some questions and answers. *The Language Teacher, 33*(3), 3-8.

Wray, A. (2002). *Formulaic language and the lexicon*. Cambridge University Press. https://doi.org/10.1017/CBO9780511519772

Author index

B
Beaupoil-Hourdel, Pauline v, 109
Belan, Sophie vi, 171
Birch-Bécaas, Susan vi, 51

D
Douglas, Dan vi, xv

F
Franklin-Landi, Rebecca vi, 153
Fries, Marie-Hélène vii, 93

H
Hoskins, Laüra vii, 51

J
Johnson, Jane Helen vii, 133
Josse, Hélène vii, 109

K
Kosmala, Loulou viii, 109

L
Labetoulle, Aude viii, 31
Le Cor, Gwen viii, 73
Lê Ngo, Thi Phuong viii, 171

M
Masuga, Katy viii, 109
Milosevic, Danica ix, 15
Morgenstern, Aliyah ix, 109

R
Rialland, Guillaume ix, 171

S
Sarré, Cédric v, 1
Schug, Daniel ix, 73
Starkey-Perret, Rebecca x, 171

W
Whyte, Shona v, 1